AF838147

Adjustment by Least Squares in Geodesy and Photogrammetry

Adjustment by Least Squares in Geodesy and Photogrammetry

R. A. Hirvonen

Professor of Geodesy,
Helsinki Technical University, Finland

FREDERICK UNGAR PUBLISHING CO., New York

Translated and revised by the author
from the original 1965 Finnish edition, TASOITUSLASKU

Second Printing, 1979

Copyright © 1971 by Frederick Ungar Publishing Co., Inc.
Printed in the United States of America
Library of Congress Catalog Card No. 71–158408
ISBN 0–8044–4397–1

Preface

This textbook grew out of lectures presented at the Helsinki Technical University, Finland, and at The Ohio State University, Columbus, Ohio. Consequently, its structure is rather pedagogic, its main purpose being the practical applications of the principle of least squares. The needs of the various groups of students suggested a division into two courses. In the elementary course, the subject is treated in full detail by the aid of many smaller problems; the most important routine adjustments of geodetic networks are included here. For the fuller understanding of the practical procedures the elementary methods of computation by the use of slide rules and desk calculators have been described. The advanced course begins with a short summary of mathematical statistics and matrix algebra, and the numerical applications are more extensive, indicating the necessity of the automatic computors in the adjustment of large nets. Some of the applications described here have been developed by the author himself and are rather new in the professional literature.

R. A. Hirvonen

Contents

I ELEMENTARY COURSE

1 INTRODUCTION

1.1 *The principal purpose of adjustments*

An adjustment is required, when, for the determination of some unknown quantities, there are more observations or measured quantities than which would be necessary if some theoretical idealizations were valid. As an illustration, we take some typical examples. For the subsequent discussion, they have been divided in two groups.

Group A:

1. There is only one unknown quantity (a distance, an angle, a height difference) which can be measured directly. Usually it is measured several times, although, according to a rough theoretical idealization, we expect that each measurement should give the same result.

2. Into a triangulation net a new point should be connected. For this purpose, two quantities should be determined, *viz.* the coordinates x and y. Usually they are not measured directly, but computed theoretically by the aid of two other quantities which can be measured directly. As the theoretical idealization, we assume that the points of the net are situated on a plane, on a sphere, or on an ellipsoid. As the quantities to be measured, we may use distance and bearing from one given point (radial measurement), bearings from two given points (intersection), two distances (section of arcs), or two angles at the new point (resection). Usually, however, more than two of these various quantities are measured.

3. For the external orientation of a photogrammetric plate six quantities are required. They can be computed, if the plate coordinates x' and y' of three points are measured for which

the ground coordinates x, y and z are known; the inner orientation of the camera and several other pertinent quantities must be known too. In a similar way, for the determination of an idealized orbit of a planet or of a satellite six quantities are required, and those can be computed if declination and right ascension are measured at three points of the orbit. Usually, in both problems, more than three points are chosen.

Group B:

4. In a triangle all three angles are measured, although two would be sufficient, since the sum of angles can be computed theoretically. When the theoretical sum is subtracted from the sum of the measured angles, there appears a small difference which is called the *misclosure*.

5. A traverse consists of a chain of sides which starts from a fixed point and a fixed direction. If the length of each side and the angle at the initial end of each side have been measured, the coordinates of the new points can be computed by the aid of successive radial measurements. Usually, the chain is "closed" into another fixed point and direction. In this case, three additional quantities are measured, *viz.* the length of the last side and the angles at both ends of it. Then the computation reveals three misclosures, *viz.* one in the end direction and one in each coordinate of the end point.

6. A levelling net consists of several closed circuits. Around each circuit, the sum of the measured height differences should be = 0; but in the practice, a small misclosure appears. The final heights of all new bench marks can be considered as the unknown quantities and the differences of each two successive bench marks as measured quantities. It can be easily shown that *the number of misclosures is equal to the difference of the numbers of the measured quantities and the unknown quantities.*

The common feature of all these examples is the *usage of the redundant measurements.* The additional measurements are always necessary because one blunder in a small number of measurements would be disastrous. Besides, it is hoped that the unavoidable small errors to some extent will compensate each other. The principal purpose of adjustments is to take advantage of this possibility of compensation which, in the practice, has proven to be very profitable. In any case, we need such a set of numerical values (for all pertinent quantities) which is consistent with the presumed theoretical idealizations.

There are two different approaches to the problem of adjustments, and we have therefore already divided our examples into groups A and B. In group A, the attention is directed first to the unknown quantities which we shall call *parameters*. These parameters must be independent of each other in such a way that the numerical value of any could not be computed from the others on the ground of the accepted theory. Instead, all the other pertinent quantities can be computed from these parameters, expecially all those quantities which have been measured directly. The parameters can often be selected in several different ways, but a constant least number u of independent parameters is — as soon as a suitable theory has been accepted — always necessary for the computation of all measurered quantities. The number n of the measured quantities is always greater than u. The difference $r = n-u$ could be called the *number of the extra measurements*, for we could select u from the measured quantities themselves as parameters, from which the rest of the measured quantities could be computed.

In group B, the attention is directed first to the *misclosures* which can be computed on the ground of the accepted theory using only the measured quantities, *i.e.*, before the parameters have been selected. Also the misclosures must be independent of each other. They could be selected in several different ways but the number of independent misclosures is always $r = n-u$.

As a matter of fact, the groups A and B have no distinction except that of the computational technique. In group A, the number u usually is smaller than r, and example 1 is the simplest case with $u=1$. In group B, the number r is smaller than u and example 4 is the simplest case with $r=1$. If in example 1 the unknown quantity had been measured two times the difference of both measured values gives one misclosure and the example could be included in group B as well.

Whether a complicated case should be included in group A or B depends, in practice, often on that which is easier: to express the measured quantities as mathematical functions of some parameters or to express the misclosures as functions of the measured quantities only. Accordingly, we need different methods of computation for each group but both methods must lead to the same results. Therefore, both methods must be derived from the same general principle.

One guiding principle can be obtained easily from the examples 1 and 4. In example 1, one quantity has been measured several times and usually we assume that all measurements are equally reliable. In that case we *intuitively* take the simple arithmetical mean of all measured values as the definitive value. In example 4 in which all three angles have been measured we usually assume that all these measurements are equally reliable. If now the sum of the measured values differs from the theoretical sum we *intuitively* divide the misclosure equally between the angles, in other words, we correct each of the measured values by the same amount. If one quantity has been measured twice, we obtain the same result whether we take the arithmetical mean of whether we halve the difference.

The guiding principle of an adjustment is that all measurements should be impartially used for the computation of the definitive results.

A contrary principle would be the *selection.* In many cases the computations are made much easier if the results of certain measurements are fixed first and not adjusted anymore while the other measurements are fitted up to them. *E.g.*, a rectangular net of stake-out lines can be checked by diagonal lines which are not used for the adjustment of the main lines in spite of small misclosures.

1.2 *The principle of least squares*

In example 1, we took the simple arithmetical mean of the

measured values as the definitive value of the parameter. Then we can compute the deviations of the individual measurements from this mean value. In a similar way, in the other examples of group A we need a general method for a simultaneous computation of all parameters after which the corresponding values of the measured quantities could be computed on the ground of the accepted theory. Then we can compute the deviations of the measured values from these "adjusted" values. The deviations are called apparent errors when computed in direction

measured value minus adjusted value

but, usually, they are computed in the other direction as corrections to be added to the measurements. We shall call them by the name

residual = adjusted value minus measured value.

In group B, on the other hand, we need a general method for the computation of the residuals, in a similar way as we did in example 4 except that in the general case all misclosures must be made to vanish simultaneously. After that, the parameters can be computed from any suitable set containing the same number of corrected measurements.

Obviously, the corrections to be added to the measured values must be as small as possible. In example 1, the sum of the residuals is = 0; thus the arithmetical mean of the residuals is as small as possible. In example 4, all residuals are equal; thus the greatest residual is as small as possible, Now we shall prove that both examples are special cases of a more general principle: *the sum of the squares of the residuals is as small as possible.*

In example 1, we denote the measured values ℓ_i $(i=1,2,\dots n)$, the unknown definitive value ℓ_a and the residuals

$$v_i = \ell_a - \ell_i$$

The square sum of these residuals is

$$[v_i v_i] = n\ell_a^2 + [\ell_i \ell_i] - 2\ell_a[\ell_i]$$

In adjustments, the brackets mean the sum of bracketed quantities or products when to the index i are given all possible values in turn; usually even the index is omitted within the brackets:

$$[vv] = n\ell_a^2 + [\ell\ell] - 2\ell_a[\ell]$$

This sum reaches it's extreme value when it's derivative with respect to the unknown ℓ_a vanishes:

$$\frac{d[vv]}{d\ell_a} = 2n\ell_a - 2[\ell] = 0$$

The solution of this equation is

$$\ell_a = \frac{[\ell]}{n}$$

which is the formula of the arithmetical mean.

In example 4, the measured values ℓ_i $(i = 1,2,3)$ give the misclosure

$$w = [\ell] - w_a$$

where $w_a = 180° +$ the spherical excess can be computed on the ground of the accepted theory. The residuals v_i must satisfy the condition

$$[v] + w = 0$$

If this condition is satisfied, the square sum $[vv]$ reaches it's extreme value for the same set of solutions as the following expression

$$S = \tfrac{1}{2}[vv] - k([v] + w)$$

where the constant k, for the present, is considered as unknown. The extreme value of S is found when all partial derivatives of it with respect to variables v_i are written = 0:

$$\frac{\partial S}{\partial v_i} = v_i - k = 0$$

This equation means that all residuals v_i are equal which was to be proved.

The principle of the least sum of squares of the residuals,

$$[vv] = \text{minimum}$$

gives the simplest known mathematical method which can be applied to any complicated case of adjustments and which, at the same time, in examples 1 and 4 gives exatly those solutions which we have chosen intuitively. Instead of it, *e.g.*, the principle of the least sum of absolute values would not be applicable to example 4 because the misclosure could be divided in many ways if only the residuals would all have the same sign; in example 1 all solutions in which the number of the positive residuals equals to that of negative residuals would be acceptable if the arithmetic mean is such a solution. Besides, the theory of the absolute values would be mathematically very complicated because it would lead to discontinuous functions. Especially, the derivation of formulas for the propagation of the errors would be very cumbersome.

Therefore, the principle of the least sum of squares of residuals has been generally accepted for the adjustments. The principle was published first by *Legendre* in 1806 with the name *Method of the least squares*. However, *Gauss* had used the same principle already since 1794 and he published several theoretical discourses and practical applications of it in 1809 and later. The method was introduced first for the computation of the orbits of planets and comets but later Gauss and *Bessel* applied it to extensive adjustments of various geodetic measurements.

The principle of least squares sometimes leads, however, to long and complicated numerical computations. On account of the practical and economical interests, we must compromise it with a second principle: the computations must be as simple and mechanical as possible. The principle of least squares is no sanction as to the theoretical preference of an adjustment; the choice of the most appropriate method often requires a personal judgment and much experience. The general trend, however, is to agree upon some schematic methods which make a personal choice unnecessary and still produce standardized results. As examples of such methods of adjustments (which are not based on the principle of least squares) we may mention the graphical adjustments (used especially for our example 2), the routine adjustment of traverses (our example 5), and polynomial adjustments of a chain or block of photogrammetric models.

1.3 *Estimation of the accuracy*

After an adjustment, the residuals give an apprehension of small errors in the measurements. By the aid of a special mathematical analysis, we may find whether these errors can be "explained away", or whether the measurements can be made more compatible with each other on the basis of an expanded and refined theoretical idealization. Anyway, a statistical investigation can give certain numerical quantities which describe the reliability of the measurements. A mathematical formulation of the *propagation of the errors* then gives similar quantities for the parameters and for all functions of them.

This *estimation of the accuracy* of various results obtainable from the measurements is the *secondary purpose of the adjustment.* Sometimes it is the principal purpose, *e.g.* when the measurements have been executed for testing new instruments or measuring methods.

There must be an abundant number of measurements before an estimation of the accuracy has any statistical significance. In practice, each quantity which will be used later in an extensive net adjustment actually is measured several times and the arithmetical mean of the original readings is then regarded as a single observation. The number of the readings is regarded proportional to the *weight* of this observation if the readings can be regarded as equally reliable. The arithmetical mean values of different measured quantities can then have different weights. Denoting the residual of one observation by v and it's weight by p, the principle of least squares, in this case, can be expressed by a generalized formula

$$[pvv] = \text{minimum}$$

on which the definitive net adjustment must be based.

Often the original readings can not be regarded as equally

reliable. We must then have other methods for the estimation of the weights. Usually we estimate first the *precision,* called also the *internal accuracy,* of each observation and define the system of weights by numbers which are inversely proportional to the squares of precision. Anyway, some hypotheses about the reliability of different observations must be accepted *before an adjustment.* On the other hand, the residuals render material for the new estimation of the reliability *after the adjustment.* These estimates describe the *external accuracy* of the observations.

The necessity to estimate the precision or the weights before an adjustment can frustrate all expectations to find definitive solutions by the method of least squares. All depends on the experience and the scientific instinct of the person who decides upon the weights and upon the observations which should be included into the net of a simultaneous adjustment.

It is, however, not profitable to waste too much time and work to the estimation of the weights. The example discussed in Chapter 5.1 shows that if the weights once estimated later are corrected by amounts which are of the same order as the original weights, the results of the adjustment will receive very small corrections which hardly have any practical significance. Only in cases when the accuracy of some groups of measured quantities after an adjustment differs very much from the precision assumed before the adjustment, a new adjustment with improved weights might be advisable.

In this connection, a few words upon the precision of the numerical computations must be added. The costs of the measurements are usually much greater than those of the computations. Therefore, it pays to carry out the computations with an essentially greater precision than the measurements. *E.g.,* the first order triangulations give usually six correct figures which means that the errors at distances of 30 km amount to a few centimeters. The computations, however, are carried out using eight figures or maintaining the millimeters. In this way, the possibility can be excluded that the accumulation of the rounding-off errors of the computation deteriorates the precision of the measurements.

In practice, the following rule of thumb is used: write down that decimal place which already is uncertain and then one more. This rule can be applied each time the arithmetic mean of several readings is computed. During the later computations, the propagation of the errors must be allowed for and the corresponding number of significant decimals maintained. As a second rule of thumb follows that the misclosures usually contain two or three significant figures. Because the weights are uncertain and less important, they may be rounded off to one or two significant figures.

1.4 *Classification of errors*

We imagine that the quantity which will be measured has a certain *true value* which, in practice, usually is unknown. The difference measured value minus the true value is called the *true error*. The classical theory of errors is based on these true errors but, nevertheless, the practical estimation of the accuracy must be based on the residuals or on the misclosures. Certainly, the misclosures can be regarded as accumulations of the true errors but only with the presumption that the mathematical theory on the ground of which the misclosures have been computed exactly corresponds to the reality.

Customarily, the errors of a measurement are divided in three classes: *blunders*, *systematic errors* and *random errors*. These words mean certain types of errors which sometimes can be recognized easily in practice but it is impossible to draw distinct boundaries between these classes. If the classification is based on the causes of the errors, we could distinguish again three different types: the errors caused by *persons*, *instruments* and *outer circumstances*.

The *blunders* are caused by the observers and by the aiding personnel. As examples, we mention figures which have been read or recorded incorrectly, wrong sign, wrong unit of measure or wrong place of the decimal point, sighting to a wrong point, carelessness in mounting or cleaning of the instruments, unnoticed jerks of the tape or the theodolite, gusts of the wind and sudden changes of the temperature. In fact, each measurement must be repeated once or several times just in order to disclose the blunders, and each repetition should be as independent of the others as possible.

If a blunder has been disclosed already during the measurements, the reading can easily be renewed, but often the blunder is discovered first during computations in the office. An expert can sometimes conclude which of the readings is faulty; in a few cases he can even guess which would be the correct reading but generally such a reading must be rejected. An "adjustment" of blunders would be quite meaningless.

The rejection of small blunders has always been a very controversial question. Some experts use a tight sieve and reject every measurement which would not fit to the others within prescribed limits or which alone causes several large misclosures. Other experts utilise conscientiously all measurements if the observer has not recorded any questionable incidents. The former method can be defended for the principal purpose of adjustments: determination of the most consistent values of all unknown quantities, the latter method for the secondary purpose: the estimation of the accuracy of these measurements.

If the observer, while still staying on the site of measurements, is aware that the range of readings is exceptionally large he can immediately make more repetitions. Definitive rejections are not recommendable in this instance. If one reading is rejected on some grounds, the surest way is to reject another reading from the opposite end of the range and make two new repetitions. The questionable readings may be useful afterwards for the statistical investigations.

The blunders can be avoided by carefulness and by rejections but, nevertheless, there always remain so called *unavoidable errors*. As an example, we suppose that the readings have been made by aid of a micrometer screw which consists of a scale and of a hair line to be brought into coincidence with the sighted point. The readings will always be scattered by various causes; due to the observer, the instrument and the external circumstances. Some of these causes are *stationary*. In a few cases, such an error can be eliminated using two readings in which the error, according to theoretical considerations, works in different directions. Usually the difference of two measurements which are affected by the same constant error is taken as an observation to be used in adjustments. Some causes do not have a constant effect but still work always in the same direction like the bends of the tape or the slant of the levelling rod. Such an *one-sided* error can not be eliminated or reduced by repetitions. In the scale of the screw, there often are *periodical errors* which can be elimated taking many readings which cover evenly whole the period.

The errors mentioned above are all examples of *systematic errors*. Usually they are caused by the instrument, often also by the external circumstances, but, in some cases, the *personal errors* can be of a systematical kind too. *E.g.*, one observer takes a reading more from the left hand side, another from the right hand side. The visual estimate of the fraction of a scale interval often is systematically wrong by some observers. *E.g.*, a statistical analysis may reveal that the readings 0.0 and 0.5 appear too often or too scantily, 0.3 and 0.7 more often than 0.2 and 0.8 and so on.

A statistical analysis can reveal systematical errors which are smaller than the random errors. In fact, there hardly is any other definition for the systematical errors than that they can be revealed afterwards by a scrutiny of the residuals. In some cases, the correlation between these errors and some probable causes can be found and numerically determined by special adjustments.

On the other hand, for the *random errors* there is no other definition than no correlation between the residuals and any possible causes, including the order of the residuals, can be found. *E.g.*, if in one reading there appears to be a positive residual, there is no way to conclude whether the next reading will have a positive or a negative residual.

Therefore, it is customary to state that the random errors are positive and negative with the same *probability*. Furthermore, experience shows that small errors have a greater frequency or probability than large errors. The ideal form of these rules is the *Gaussian law of errors*, called also the *normal distribution*, which is dealt with in Chapter 11.3. This empirical law has, however, one theoretical foundation: if several different systematical errors are present in the measurements and no of those errors is dominating, then the sum of a great number of systematical errors tends towards the normal distribution. Therefore, in practice it usually is impossible to distinguish all systematic errors from each other and from the possible "true" random errors.

Sometimes, such a statement has been put forth that an adjustment by the method of least squares could be used only under the presumption that the errors involved, *i.e.* the residuals, are of random character. Of course, such a presumption can never be taken for granted before the adjustment has been carried through and the residuals analysed in all possible ways. For purely practical purposes, anyway, an adjustment by least squares often is the simplest and most reliable method to ascertain consistent values for the unknown quantities from redundant measurements. If the residuals show a random character, the values obtained have the greatest *probability* of all possible solutions. But this does not mean very much in practice, if we remember the uncertainty of the weights which had to be estimated beforehand. In fact, the statement mentioned above is valid only in that respect that the estimation of the accuracy is reliable under the said presumption. In other words, the random character of errors is necessary for the secondary purpose of the adjustment, not for the principal purpose as defined in Chapter 1.1.

2 FUNDAMENTAL CONCEPTS

2.1 *Measures of precision*

Suppose that a quantity x is measured n times by the same method and in the same circumstances. The measurements give slightly different values x_i $(i = 1,2,...n)$ and the *arithmetic mean*

$$x_m = \frac{[x]}{n} \tag{2.01}$$

is regarded as the definitive result of the measurements.

In our theoretical discussions, we regard x as a *stochastic variable*. It implies the presumption that if the measurements could be repeated infinitely in the same circumstances, x_m would tend to a certain limit

$$x_t = \lim_{n\to\infty} x_m \tag{2.02}$$

which is called the *theoretical mean*. In the classical theory of errors, the corresponding idealization of x was called the "true value". Anyway, x_t is an unknown quantity and all we can do in the practice is to use the arithmetical mean x_m as an *estimate* of x_t.

The mathematical theory of errors is based on the "true errors" or theoretical errors

$$\varepsilon_i = x_i - x_t \tag{2.03}$$

but our practical computations must be based on the "apparent errors" $x_i - x_m$. In order to avoid the name "errors", however, we shall take these differences in the opposite direction like "corrections" and call them *residuals*:

$$v_i = x_m - x_i \tag{2.04}$$

Anyway, the signs of the errors are usually unimportant. When we speak about the maximum error we mean the error with the greatest absolute value. When necessary, the sign can be easily allowed for. *E.g.*, the theoretical error of the arithmetical mean is obviously

$$x_m - x_t = v_i + \varepsilon_i \tag{2.05}$$

The measures of precision are defined by the theoretical errors. Therefore, the numerical values of these measures remain usually unknown and we must introduce acceptable definitions for the *estimates* of these measures. We may require that each estimate tends to it's theoretical counterpart when n increases infinitely. In such a case, the estimate is called *unbiased.*

1. *The maximum error* is the greatest absolute value which can be expected to occur in an infinite number of theoretical errors ε. It can be estimated by the greatest of the n available residuals. Of course, if n is a small number, this estimate is not very reliable.

As an advantage of the maximum error, it has a very simple *law of propagation*. *E.g.*, if a quantity x has maximum error α and another quantity y has maximum error β, both the sum $x+y$ and the difference $x-y$ have the maximum error $\alpha+\beta$. In practice, however, this law is rather "pessimistic". If we take at random two values x_i and y_i, it certainly is possible but very improbable that they both have the greatest errors with due signs.

In practice, the *maximum permitted error* is often used. It means that each series of measurements which reveals residuals which exceed a given limit must be rejected or remeasured.

2. *The probable error* r will be defined theoretically in Chapter 11.3. In practice, it can be estimated arranging the absolute values $|v|$ of the residuals according to their magnitude and picking up the value at the middle.

In spite of this simple method of estimation, the probable error has certain theoretical defects. It does not reflect the effect of the greatest residuals which usually are decisive for an estimation of the reliability. The law or error propagation could not be found or used easily.

3. *The average error* or the *mean of errors* is usually estimated by formula

$$t = \frac{[|v|]}{n}$$

Theoretically, it could be defined by formula

$$\tau = \lim_{n\to\infty} \frac{[|\varepsilon|]}{n} \tag{2.06}$$

The former formula, however, does not give an unbiased estimate of the latter, for if we have g groups of measurements with n single measurements in each group, the mean value of all estimates t would not tend to τ when g increases infinitely. The estimate t, however, is used much because it can be obtained without extensive computations in the connection of the checking of the mean value x_m. For the latter purpose, we usually sum up all positive values of v and then all negative values. If the former sum is s, the latter should be $-s$ because

$$[v] = 0 \tag{2.07}$$

Now we obtain simply

$$t = \frac{2s}{n} \tag{2.08}$$

The crucial disadvantage of the average error is that the algebraic formulas for the propagation of errors would be cumbersome.

4. Therefore, the theoretical mean of the squares of errors

$$\sigma^2 = \lim_{n\to\infty} \frac{[\varepsilon\varepsilon]}{n} \tag{2.09}$$

has been accepted as the fundamental measure of the precision. It is called the *variance* or the theoretical variance. As it is always positive, it has a real square root σ which is called *mean error* or root mean square error (rms-error).

In practice, we need a formula for the unbiased estimate of the mean error. This formula must be based on the utilization of the residuals v and in it will appear the number n too.

The theoretical requirement for that m were an unbiased estimate of σ is:

$$\sigma = \lim_{g\to\infty} \frac{[m]}{g} \qquad (2.10)$$

where g is the number of groups of measurements with n single measurements in each group. Such an estimate will be derived later, see formula (4.11). According to the established terminology of adjustments, we shall call the unbiased estimate of rms-error by the name *standard error*.

Meanwhile, we give formulas for other estimates which are not unbiased:

$$m_t = \frac{5}{4}\,t \qquad (2.11)$$

$$m_r = \frac{3}{2}\,r \qquad (2.12)$$

$$m_v = \sqrt{\frac{|vv|}{n}} \qquad (2.13)$$

The formulas (2.11) and (2.12) are based on the presumption of the normal distribution as explained in Chapter 11.3. In practice, some other types of distribution can be met. We shall take two very common examples.

The *rounding-off error* has an *even distribution* over the range $-0.5 < \varepsilon < +0.5$ units of the last decimal place which is written down. It's variance is, according to formula (2.09):

$$\sigma^2 = \int_{-0.5}^{+0.5} \varepsilon^2 d\varepsilon = \frac{1}{12}$$

and the mean error

$$\sigma = \frac{\sqrt{3}}{6} = \pm\ 0.29$$

The probable error and the average error are both

$$r = \tau = 0.25$$

A *periodical error* $\varepsilon = k \sin \alpha$ appears in geodetic measurements which are made with graduated scales or micrometer screws;

it is very common in some astronomical or geophysical observations. Supposing that the angle α has a uniform distribution around the whole period, the probable error is

$$r = k \sin \frac{\pi}{4} = 0.707\ k$$

By the aid of integrations, we obtain for the average error

$$\tau = \frac{2k}{\pi} = 0.647\ k$$

and for the mean error $\sigma = r$.

2.2 *The covariance*

One series of measurements gives sometimes a result which must be expressed by two distinct quantities x and y as if it were a gun hit on the target. Repeated series give a *two-dimensional stochastic variable* with components x_i and y_i $(i=1,2,\ldots n)$.

Each component can be considered also as a common one-dimensional variable with variances

$$\sigma_x{}^2 = \lim_{n\to\infty} \frac{[\varepsilon_x \varepsilon_x]}{n} \quad \text{and} \quad \sigma_y{}^2 = \lim_{n\to\infty} \frac{[\varepsilon_y \varepsilon_y]}{n} \tag{2.14}$$

In addition, we need a numerical measure for the mutual correlation of the components which expresses how strongly they "depend on each other." For this purpose, we define a new quantity

$$\sigma_{xy} = \lim_{n\to\infty} \frac{[\varepsilon_x \varepsilon_y]}{n} \tag{2.15}$$

which is called *covariance*. It can be positive or negative; therefore it's square root has no real significance like the mean errors σ_x and σ_y. If x and y are quite independent of each other, the covariance is = 0. On the other hand, we can not conclude that x and y are mathematically independent of each other when the covariance is zero but we say that they are *uncorrelated*.

As an example, we imagine that the gun is aimed in the direction of x-axis of a horizontal target area. Then, usually, the dispersion of the hits will be greater in this direction than in the direction of y-axis. Now an unaccounted wind from the left causes a correlation: if the shot is too long, the transversal deviation goes generally to the right and vice versa. The hits are scattered over an area which can be idealized by an ellipse the axes of which are not parallel with x- and y-axes.

The absolute value of the covariance cannot be greater than the product of both mean errors:

$$|\sigma_{xy}| \leqq \sigma_x \sigma_y$$

The ratio

$$\rho_{xy} = \frac{\sigma_{xy}}{\sigma_x \sigma_y} \tag{2.16}$$

which is called *coefficient of correlation* is used very much in mathematical statistics. The extreme values of $|\rho_{xy}|$ are 0 and 1. The former means that x and y are uncorrelated, the latter means that they are exact linear functions of each other. In the former case, the axes of the ellipse of dispersion are parallel with x- and y-axes, in the latter case, the ellipse is degenerated to a segment of straight line.

A *multidimensional stochastic variable* has several components, x, y, z..., which all have their own variances, σ_x^2, σ_y^2, σ_z^2..., and a great number of covariances, σ_{xy}, σ_{xz}, σ_{yz}... If the number of dimensions is ν the number of covariances is $\frac{1}{2}\nu(\nu-1)$.

3 *THE PROPAGATION OF ERRORS*

3.1 *Fundamental formulas*

Let x be a stochastic variable with the theoretical mean x_t and with mean error σ_x. Let $u = u(x)$ be a mathematical function of x. Hence, u is also a stochastic variable. The theoretical mean of u is usually supposed to be

$$u_t = u(x_t) \tag{3.01}$$

To be exact, this is true only for linear functions. In the practice, we assume that the errors ε of x are small and that the errors δ of u can be, with a sufficient precision, expressed by formula

$$\delta = \varepsilon \frac{du}{dx} \tag{3.02}$$

in which the value of the derivative is constant:

$$\frac{du}{dx} = u'(x_t) \tag{3.03}$$

With the same presumptions, we obtain from the definition of the mean error

$$\sigma_u = \sigma_x \frac{du}{dx} \tag{3.04}$$

In the function u = u(x) is an adequate mathematical model of the empirical correlation between two measurable quantities x and u, then we can expect that the repeated measurements give results which, within certain limits of the precision, comply with equations

$$u_m = u(x_m)$$
$$m_u = m_x \frac{du}{dx}$$

If these formulas seem to be fallacious in practice, in spite of the sufficient precision of the measurements, we may conclude that the mathematical model is not adequate.

Let now x,y be a two-dimensional stochastic variable and $u = u(x,y)$ a mathematical function representing a one-dimensional variable. With similar presumptions as above, we can write

$$u_t = u(x_t, y_t) \tag{3.05}$$

$$\delta_u = \varepsilon_x \frac{\partial u}{\partial x} + \varepsilon_y \frac{\partial u}{\partial y} \tag{3.06}$$

The definitions of the variance and covariance gives now

$$\sigma_u^2 = \sigma_x^2 \left(\frac{\partial u}{\partial x}\right)^2 + \sigma_y^2 \left(\frac{\partial u}{\partial y}\right)^2 + 2\sigma_{xy} \frac{\partial u}{\partial x} \frac{\partial u}{\partial y} \tag{3.07}$$

where the last term disappears if x and y are uncorrelated.

Let $v = v(x,y)$ be another mathematical function of x,y. Then u,v is a new two-dimensional stochastic variable. The component v has a variance

$$\sigma_v^2 = \sigma_x^2 \left(\frac{\partial v}{\partial x}\right)^2 + \sigma_y^2 \left(\frac{\partial v}{\partial y}\right)^2 + 2\sigma_{xy} \frac{\partial v}{\partial x} \frac{\partial v}{\partial y} \tag{3.08}$$

The covariance of u and v is

$$\sigma_{uv} = \sigma_x^2 \frac{\partial u}{\partial x} \frac{\partial v}{\partial x} + \sigma_y^2 \frac{\partial u}{\partial y} \frac{\partial v}{\partial y} + \sigma_{xy}\left(\frac{\partial u}{\partial x} \frac{\partial v}{\partial y} + \frac{\partial u}{\partial y} \frac{\partial v}{\partial x}\right) \tag{3.09}$$

Note that σ_{uv} is not zero even if x and y are uncorrelated in which case only the last term disappears. In the general case, two stochastic variables are correlated with each other if they can be expressed as mathematical functions of one or more common variables. In other words: two different observed quantities are correlated if they have common sources of variations. This is the theoretical definition of the systematical errors.

The formulas given above can be generalized for the case in which the number of both the original variables and the functions is more than two. In order to give all the necessary formulas in form of one equation, we write σ_{xx} instead of σ_x^2, σ_{uu} instead of σ_u^2 *etc.* Then the general formula reads:

$$\sigma_{uv} = \sigma_{xx}\frac{\partial u}{\partial x}\frac{\partial v}{\partial x} + \sigma_{xy}\frac{\partial u}{\partial x}\frac{\partial v}{\partial y} + \sigma_{xz}\frac{\partial u}{\partial x}\frac{\partial v}{\partial z} + \cdots$$
$$+ \sigma_{xy}\frac{\partial u}{\partial y}\frac{\partial v}{\partial x} + \sigma_{yy}\frac{\partial u}{\partial y}\frac{\partial v}{\partial y} + \sigma_{yz}\frac{\partial u}{\partial y}\frac{\partial v}{\partial z} + \cdots \qquad (3.10)$$
$$+ \sigma_{xz}\frac{\partial u}{\partial z}\frac{\partial v}{\partial x} + \sigma_{yz}\frac{\partial u}{\partial z}\frac{\partial v}{\partial y} + \sigma_{zz}\frac{\partial u}{\partial z}\frac{\partial v}{\partial z} + \cdots$$

Formula (12.53) shows how this formula can be written in a very concise form by the aid of matrix notations.

If the original variables $x, y, z \ldots,$ are all uncorrelated in all possible combinations, then $\sigma_{xy} = \sigma_{xz} = \sigma_{yz} = \cdots = 0$. In this case, we obtain "the special law of error propagation":

$$\sigma_u = \sqrt{\left(\sigma_x \frac{\partial u}{\partial x}\right)^2 + \left(\sigma_y \frac{\partial u}{\partial y}\right)^2 + \left(\sigma_z \frac{\partial u}{\partial z}\right)^2 + \cdots} \qquad (3.11)$$

and the special formula for the covariance of two functions

$$\sigma_{uv} = \sigma_x^2 \frac{\partial u}{\partial x}\frac{\partial v}{\partial x} + \sigma_y^2 \frac{\partial u}{\partial y}\frac{\partial v}{\partial y} + \sigma_z^2 \frac{\partial u}{\partial z}\frac{\partial v}{\partial z} + \cdots \qquad (3.12)$$

In Chapter 14.2 we return to the inverse problem: how to find uncorrelated functions of correlated variables.

In practical applications, the theoretical variances and covariances usually are unknown. Therefore, we must replace the symbols σ in all formulas given above by the symbols m which mean the unbiased estimates of the corresponding theoretical values.

3.2 *General applications*

If the standard errors of x and y are m_x and m_y, respectively, and if the covariance of x and y has been estimated by m_{xy}, then:

The standard error of the sum $s = x+y$ is

$$m_s = \sqrt{m_x^2 + m_y^2 + 2m_{xy}} \qquad (3.13)$$

The standard error of the difference $d = x-y$ is

$$m_d = \sqrt{m_x^2 + m_y^2 - 2m_{xy}} \qquad (3.14)$$

The covariance of the sum and the difference is

$$m_{sd} = m_x{}^2 - m_y{}^2 \tag{3.15}$$

In the special case that x and y are uncorrelated their sum and difference have both the same standard error

$$m_s = m_d = \sqrt{m_x{}^2 + m_y{}^2} \tag{3.16}$$

Formula (3.13) and it's special case (3.16) can easily be expanded as to be valid for the sum of more than two quantities. This formula for the error propagation of a sum is valid also in the case when a measured quantity is variable because of two or more sources of errors and when these errors can be estimated by the aid of variances and possible covariances.

If the sum contains n terms, or if there are n sources of errors, which all can be characterized by the same standard error m without any covariances, the sum, or the measurement, has a combined standard error

$$m_n = m\sqrt{n} \tag{3.17}$$

This formula appears very often in geodetic measurements. *E.g.*, if a long distance has been measured with a tape and if the measurement of each tape length has the same standard error m, the individual errors being independent of each other, then the standard error of the entire distance of n tape lengths is $m\sqrt{n}$. The same formula is supposed to be valid in a spirit levelling when the sight distances are equal. In other words, the standard error of a tape measurement and that of a spirit levelling are supposed to be proportional to the square root of the entire distance.

To be sure, we note that formula (3.17) is valid only when the individual terms, or the sources of errors, are independent of each other. *E.g.*, if there is a constant error ε in the length of a tape, this error will propagate according to formula $n \cdot \varepsilon$. One-sided errors, *e.g.* those caused by the curves and slopes of the tape, even if they can be estimated statistically by a standard error m, propagate similarly according to formula

$n \cdot m$.

If a quantity is multiplied or divided by a constant, the standard error of the quantity must be multiplied or divided, respectively, by the same constant. According to this rule, we get from (3.17) which gives the standard error of the sum of n similar observations, the standard error of the arithmetical mean:

$$M = \frac{m}{\sqrt{n}} \tag{3.18}$$

when m is the standard error of one observation. This formula seems to indicate that the precision of the mean value could be improved infinitely by repetition of the individual measurements. Indeed, it is advisable to repeat several times all measurements, but the decrease of the standard error of the mean will eventually slow down and stop at a limit where the constant or one-sided errors attain the dominating effect.

According to (3.07), *the standard error of the product* $t = x \cdot y$ is

$$m_t = \sqrt{(y \cdot m_x)^2 + (x \cdot m_y)^2 + 2xy \cdot m_{xy}} \tag{3.19}$$

In the special case that x and y are uncorrelated we obtain

$$m_t = \sqrt{(y \cdot m_x)^2 + (x \cdot m_y)^2} \tag{3.20}$$

Dividing both sides of (3.19) by $t = xy$, we obtain

$$\frac{m_t}{t} = \sqrt{\left(\frac{m_x}{x}\right)^2 + \left(\frac{m_y}{y}\right)^2 + 2\,\frac{m_{xy}}{xy}} \tag{3.21}$$

In this formula, there appears three times an expression of type: standard error of a quantity divided by the quantity itself. This dimensionless number is called *relative standard error*. Formula (3.21) can easily be expanded as to be valid for the product of more than two quantities. Of course, all possible covariances of the factors must be allowed for.

The standard error of the quotient $q = x/y$ is

$$m_q = \sqrt{\left(\frac{m_x}{y}\right)^2 + \left(\frac{x \cdot m_y}{y^2}\right)^2 - \frac{2x}{y^3}\,m_{xy}} \tag{3.22}$$

Dividing both sides by $q = x/y$, we obtain

$$\frac{m_q}{q} = \sqrt{\left(\frac{m_x}{x}\right)^2 + \left(\frac{m_y}{y}\right)^2 - 2\frac{m_{xy}}{xy}} \tag{3.23}$$

The standard error of the natural logarithm $u = \ln x$ is

$$m_u = \frac{m_x}{x} \tag{3.24}$$

or the same as the relative standard error of the number itself.

The standard error of the common logarithm $u = \log x$ is

$$m_u = \mu \frac{m_x}{x} \qquad \text{where} \qquad \mu = 0.4343 \tag{3.25}$$

If a mathematical function of the multidimensional variable can be divided in factors in the following way:

$$u = f(x) \cdot g(y) \tag{3.26}$$

we compute the relative standard error of the function by the aid of the natural logarithms

$$\ln u = \ln f(x) + \ln g(y) \tag{3.27}$$

Differentiating (3.27), we obtain first

$$\frac{du}{u} = \frac{f'}{f}\,dx + \frac{g'}{g}\,dy \tag{3.28}$$

and then the computation formula for the standard error gives

$$\frac{m_u}{u} = \sqrt{\left(\frac{f'}{f}m_x\right)^2 + \left(\frac{g'}{g}m_y\right)^2 + 2\frac{f'g'}{fg}m_{xy}} \tag{3.29}$$

The relative standard errors can be given in percents, or, in form $1:K$ where K is a great number. Sometimes, it is advisable to multiply the relative standard errors of the sides of a trilateration by constant $\rho'' = 206265''$ or $\rho'' = 636620''$. In this way, the precision of the side measurements can be expressed in (sexagesimal or centesimal) seconds of arc like the angle measurements in a triangulation.

3.3 *Geodetic applications and examples*

1. Suppose that the random standard error of a *base line measurement* is $m = \pm 0.03$ mm for each tape length of 24 metres. If now a standardization base of 480 m (or 20 tape lengths) has been measured with one tape, the standard error of the measurement is, according to (3.17), $0.03 \cdot \sqrt{20} = 0.134$ mm. The length of the base being known, the measurement gives the length of the tape with a standard error of $0.134/20 = 0.0067$ mm. When now another base is measured with this tape, there are two different types of errors: the same random error as above and the systematic error of the tape. If the new base is 720 m long (or 30 tape lengths), the random errors accumulate to $0.03 \cdot \sqrt{30} = 0.164$ mm and the systematic errors to $30 \cdot 0.0067 = 0.201$ mm. The former being independent of the latter, the combined effect of both errors is, according to (3.16),

$$\sqrt{0.164^2 + 0.201^2} = \pm 0.26 \text{ mm}$$

2. The principal formulas for the *computation of plane coordinates* read

$$\begin{aligned} x &= s \cos t \\ y &= s \sin t \end{aligned} \qquad (3.30)$$

where s is the measured side, t the bearing and x, y the coordinate differences to be computed. Partial differentiations give

$$\begin{aligned} \frac{\partial x}{\partial s} &= \cos t = \frac{x}{s}, & \frac{\partial x}{\partial t} &= -s \sin t = -y \\ \frac{\partial y}{\partial s} &= \sin t = \frac{y}{s}, & \frac{\partial y}{\partial t} &= +s \cos t = +x \end{aligned} \qquad (3.31)$$

Here the unit of the angles is radian. If the standard error of the bearing is m_t seconds of arc, that of the side m_s, and the covariance $m_{st} = 0$, we obtain from (3.11)

$$m_x = \sqrt{\left(\frac{x}{s}m_s\right)^2 + \left(\frac{y}{\rho}m_t\right)^2}$$

$$m_y = \sqrt{\left(\frac{y}{s}m_s\right)^2 + \left(\frac{x}{\rho}m_t\right)^2} \tag{3.32}$$

and from (3.12)

$$m_{xy} = xy\left\{\left(\frac{m_s}{s}\right)^2 - \left(\frac{m_t}{\rho}\right)^2\right\} \tag{3.33}$$

As an example, let be $s = 215.640$ metres, $m_s = \pm 60$ mm, $t = 62°26'15"$, $m_t = \pm 45"$. The principal computation gives $x = 99.780$ m, $y = 191.166$ m. If we first write down the intermediate results $m_s/s = 278$ mm/km, $m_t/\rho = 218$ mm/km, (3.32) and (3.33) give easily

$$m_x = \sqrt{(0.0998 \cdot 278)^2 + (0.1912 \cdot 218)^2} = \pm 50 \text{ mm}$$

$$m_y = \sqrt{(0.1912 \cdot 278)^2 + (0.0998 \cdot 218)^2} = \pm 57 \text{ mm}$$

$$m_{xy} = +\ 569 \text{ mm}^2$$

The coefficient of correlation is $\rho_{xy} = 569/2850 = +\ 0.198$.

3. *The inverse formulas of the coordinate computation* read

$$s = \sqrt{x^2 + y^2} \tag{3.34}$$

$$t = \arctan\frac{y}{x} \quad \text{when} \quad |y| < |x|$$

$$t = \operatorname{arccot}\frac{x}{y} \quad \text{when} \quad |x| < |y| \tag{3.35}$$

The partial derivatives can be formed most easily as an inverse matrix when the expressions (3.31) are taken as elements of the original matrix:

$$\frac{\partial s}{\partial x} = \frac{x}{s}, \qquad \frac{\partial s}{\partial y} = \frac{y}{s}$$

$$\frac{\partial t}{\partial x} = -\frac{\rho y}{s^2}, \qquad \frac{\partial t}{\partial y} = +\frac{\rho x}{s^2} \tag{3.36}$$

If m_x and m_y are given and if $m_{xy} = 0$, we obtain from (3.11)

$$m_s = \sqrt{(\tfrac{x}{s} m_x)^2 + (\tfrac{y}{s} m_y)^2}$$
$$m_t = \tfrac{\rho}{s}\sqrt{(\tfrac{y}{s} m_x)^2 + (\tfrac{x}{s} m_y)^2} \qquad (3.37)$$

and from (3.12)

$$m_{st} = \tfrac{\rho}{s}\,\tfrac{x}{s}\,\tfrac{y}{s}(m_y^2 - m_x^2) \qquad (3.38)$$

As an example, let x, y, m_x and m_y be equal to the resulting values of the previous example but suppose that $m_{xy} = 0$. The principal computation, of course, gives s and t as equal to the starting values of the previous example. If we write first the intermediate results $x/s = 0.463$, $y/s = 0.887$ and $\rho/s = 0\overset{''}{.}957/\text{mm}$, (3.37) and (3.38) give easily

$$m_s = \sqrt{(50\cdot 0.463)^2 + (57\cdot 0.887)^2} = \pm\ 56\ \text{mm}$$

$$m_t = 0\overset{''}{.}957\ \sqrt{(50\cdot 0.887)^2 + (57\cdot 0.463)^2} = \pm\ 49''$$

$$m_{st} = +\ 294\ \text{mm}\cdot 1''$$

The coefficient of correlation is $\rho_{st} = 294/2744 = +\ 0.107$.

In both of these examples, a geometric interpretation of the covariance can be obtained by the aid of error ellipses as explained in Chapter 14.1.

4. The fundamental formulas for the *solution of a plane triangle* read

$$\alpha + \beta + \gamma = \pi = 180^\circ = 200^g \qquad (3.39)$$

$$c \cos \alpha = b - a \cos \gamma \qquad (3.40)$$

$$c \sin \alpha = a \sin \gamma \qquad (3.41)$$

from which the computation formulas can be obtained for all cases of the general problem: given three parts from which at least one is a side, wanted one, two or all three from the remaining parts. *E.g.*, if all three sides are given, the square sum of (3.40)(3.41) gives the expanded formula of Pythagoras:

$$c^2 = a^2 + b^2 - 2ab \cos \gamma \tag{3.42}$$

which gives angle γ. First the angle should be computed which is most close to the straight angle (90°), then (3.41) can be used for the computation of smaller angles and (3.39) for the checking.

From the fundamental formulas, we can derive the differential formulas

$$d\alpha + d\beta + d\gamma = 0 \tag{3.43}$$

$$dc = da \cos \beta + db \cos \alpha + \frac{d\gamma}{\rho}(a \sin \beta = b \sin \alpha) \tag{3.44}$$

$$\frac{dc}{c} + \frac{d\alpha}{\rho} \cot \alpha = \frac{da}{a} + \frac{d\gamma}{\rho} \cot \gamma \tag{3.45}$$

$$da \sin \beta = db \sin \alpha + \frac{c}{\rho}\, d\alpha + \frac{a}{\rho}\, d\gamma \cos \beta \tag{3.46}$$

When three parts are given and one part is wanted, we can always find from formulas (3.44)–(3.46) such one where the differentials of these four parts appear. Then we can form the partial derivatives which are required for the general formula (3.10). E.g., from (3.44) we obtain $\frac{\partial c}{\partial a} = \cos \beta$ but from (3.45) $\frac{\partial c}{\partial a} = \frac{c}{a}$. The former must be used when, for the computation of c, a, b and γ are given, the latter, when a, α and γ are given.

As an example, we take the trigonometric determination of a long distance. Suppose that the base b = 49.858 meters has been measured with a standard error ± 2mm and the angles $\alpha = 67°27'18''$ and $\gamma = 85°13'12''$ both with a standard error ± 5"; all three measurements are independent of each other.

As to the principal computation, the easiest way is to use the angle $\beta = 27°19'30''$ by the aid of which we obtain

$$a = b\,\frac{\sin \alpha}{\sin \beta} = 100.314 \tag{3.47}$$

For the computation of the standard error of a, however, we must use formula (3.46) displaying the differentials da, db, $d\alpha$ and $d\gamma$. With the partial derivatives

$$\frac{\partial a}{\partial b} = \frac{a}{b} = 2.012, \quad \frac{\partial a}{\partial \alpha} = \frac{c}{\rho \sin \beta} = 1.143, \quad \frac{\partial a}{\partial \gamma} = \frac{a}{\rho} \cot \beta = 0.941 \text{ mm/1"}$$

we obtain

$$m_a = \pm\sqrt{4.024^2 + 5.715^2 + 4.705^2} = \pm 8.4 \text{ mm}$$

It could be misleading to use formula (3.45) corresponding to the principal computation (3.47). Interchanging in (3.45) c and b, likewise γ and β, we would obtain

$$\frac{\partial a}{\partial b} = \frac{a}{b}, \quad \frac{\partial a}{\partial \alpha} = \frac{a}{\rho} \cot \alpha, \quad \frac{\partial a}{\partial \beta} = -\frac{a}{\rho} \cot \beta$$

In this case, we should compute first $m_\beta = 5" \cdot \sqrt{2}$ and $m_{\alpha\beta} = -(5")^2$ If the covariance is not forgotten, the correct value of m_a will be found in an alternative but more complicated way.

5. *The area of a segment* can be computed by formula

$$A = \frac{1}{2} r^2 \left(\frac{\alpha}{\rho} - \sin \alpha\right) \tag{3.48}$$

when r = the radius of the circle and α = the central angle are given. The differentiation gives

$$dA = dr\, r\left(\frac{\alpha}{\rho} - \sin \alpha\right) + \frac{1}{2} \frac{d\alpha}{\rho} r^2 (1 - \cos \alpha) \tag{3.49}$$

As an example, take $r = 62.56 \pm 0.03$ meters and $\alpha = 75°44' \pm 2'$. Then we have $\sin \alpha = 0.96916$, $\frac{\alpha}{\rho} = 1.32180$, $\cos \alpha = 0.246$, $\frac{\partial A}{\partial r} = 0.2206$, $\frac{\partial A}{\partial \alpha} = 0.429 \text{ m}^2/1'$ and, eventually, $A = 690.07$ with

$$m_A = \sqrt{0.662^2 + 0.858^2} = \pm 1.08 \text{ m}^2$$

If we compute first the area of the sector

$$S = \frac{1}{2} r^2 \frac{\alpha}{\rho} = 2586.600$$

and the area of the triangle

$$T = \frac{1}{2} r^2 \sin \alpha = 1896.527$$

with the variances

$$m_S^2 = (r \frac{\alpha}{\rho} m_r)^2 + (\frac{1}{2} r^2 \frac{m_\alpha}{\rho})^2 = 7.45$$

$$m_T^2 = (r \sin\alpha\, m_r)^2 + (\frac{1}{2} r^2 \cos\alpha \frac{m_\alpha}{\rho})^2 = 3.39$$

we must not forget the covariance

$$m_{ST} = r^2 \frac{\alpha}{\rho} \sin\alpha\, m_r^2 + \frac{1}{4} r^4 \cos\alpha (\frac{m_\alpha}{\rho})^2 = 4.83$$

Eventually, we obtain

$$A = S - T = 690.07$$

$$m_A = \sqrt{m_S^2 + m_T^2 - 2\, m_{ST}} = \pm 1.09$$

6. The fundamental formulas for the *solution of a spherical triangle* read

$$\cos c = \cos a \cos b + \sin a \sin b \cos\gamma \tag{3.50}$$
$$\sin c \cos\alpha = \cos a \sin b - \sin a \cos b \cos\gamma \tag{3.51}$$
$$\sin c \sin\alpha = \sin a \sin\gamma \tag{3.52}$$

which can be immediately used for the case where the given parts are two sides a and b and the angle γ between them. Additional formulas can be obtained by the *polarization*, *i.e.* replacing each part by the supplement of the opposite part:

$$\cos\gamma = -\cos\alpha \cos\beta + \sin\alpha \sin\beta \cos c \tag{3.53}$$
$$\sin\gamma \cos a = +\cos\alpha \sin\beta + \sin\alpha \cos\beta \cos c \tag{3.54}$$

In theory, the remaining parts of a spherical triangle can always be computed when any three parts are given. In practice, the result is not always very accurate. In some cases, the formulas given above do not allow an immediate solution. *E.g.*, when two sides a and c and the angle γ opposite to the second given side are given, we must compute first auxiliary parts m and M by formulas

$$m \cos M = \cos a \tag{3.55}$$
$$m \sin M = \sin a \cos\gamma \tag{3.56}$$

after which (3.50) gives

$$\cos (b - M) = \frac{1}{m} \cos c \tag{3.57}$$

From the fundamental formulas we obtain by differentiations

$$dc = da \cos \beta + db \cos \alpha + d\gamma \sin a \sin \beta \tag{3.58}$$

$$dc \cot c + d\alpha \cot \alpha = da \cot a + d\gamma \cot \gamma \tag{3.59}$$

$$da \sin \beta = db \cos c \sin \alpha + d\alpha \sin c + d\gamma \sin a \cos \beta \tag{3.60}$$

$$da \sin b \sin \gamma = d\alpha + d\beta \cos c + d\gamma \cos b \tag{3.61}$$

In each of these equations, there appear four differentials from which three represent the given parts and one the part to be computed. All possible combinations can be found from these four formulas. The correct formula must be used, even if one or two of the given parts remain constant, *i.e.* are supposed to be exactly known. Often only one of the unknown parts is wanted for the principal problem. For the computation of it's standard error, the other unknown parts may be required too but then a minor accuracy (*e.g.* that of a slide rule) is sufficient. Sometimes, especially when working with a high speed computer, it is better to repeat the principal computations several times, changing each time one of the given parts by the positive amount of it's standard error.

As an example, we compute the azimuth of a star when the hour angle and the declination of the star and the latitude of the observation station are given. First, the notations used in the formulas above must be replaced by those of the spherical astronomy:

Given parts: $\gamma = t$ (hour angle) $= 6^h 29^m 01^s.2 \pm 0^s.6$

$90^\circ - a = \delta$ (declination) $= 20^\circ 12' 34''.2$

$90^\circ - b = \phi$ (latitude) $= 60^\circ 09' 45''.0$

Wanted part: $180^\circ - \alpha = A$ (azimuth)

Other parts: $c = z$ (zenith distance) and $\beta = q$ (parallactic angle)

The principal computation can be carried out by formulas

$$\sin z \cos A = -\cos \phi \sin \delta + \sin \phi \cos \delta \cos t \tag{3.62}$$

$$\sin z \sin A = \cos \delta \sin t \tag{3.63}$$

which give $A = 106°26'21''.3$. For the computation of the standard error, we must use formula (3.60) which after the replacement of the notations gives

$$\frac{\partial A}{\partial t} = \frac{\cos\delta\cos q}{\sin z} = \frac{\sin A\cos q}{\sin t}$$

Here we need the parallactic angle q for which we have

$$\sin q = \frac{\cos\phi\sin A}{\cos\delta}$$

The computation gives $\frac{\partial A}{\partial t} = 0.83$. Hence, $m_A = 0.83 \cdot m_t = \pm 7''.5$. In this case, it is better to repeat the principal computation for the value $t + |m_t| = 6^h29^m01^s.8$. Then we obtain $A + |m_A| = 106°26'\ 28''.8$ and $m = \pm 7''.5$.

7. The solution of geodetic triangles and the computation of coordinates *on the ellipsoid* imply long fundamental formulas which are not given here. As to the computation of the standard errors, the differential formulas (3.43)–(3.46) or (3.58)–(3.61) often give a sufficient approximation. Here, however, we shall introduce new notations.

The principal problem, *computation of coordinates*, involves the given quantities side s, azimuth A_1 and latitude B_1 The quantities to be computed are longitude difference ℓ, latitude of the new point B_2 and the back-azimuth A_2. With the substitutions

$$a = \rho\frac{s}{r}, \qquad b = 90° - B_1, \qquad c = 90° - B_2$$
$$\alpha = \ell, \qquad \beta = 360° - A_2, \qquad \gamma = A_1$$

we obtain from (3.58) and (3.60) when $\frac{s}{r}$ and ℓ are small angles

$$dB_2 = -\, ds\,\frac{\rho}{r}\cos A_2 + dB_1 + dA_1\,\frac{s}{r}\sin A_2 \tag{3.64}$$

$$d\ell = -\, ds\,\frac{\rho}{r}\,\frac{\sin A_2}{\cos B_2} + dB_1\sin\ell\tan B_2 - dA_1\,\frac{s}{r}\,\frac{\cos A_2}{\cos B_2} \tag{3.65}$$

Interchanging first a and b, and likewise α and β, (3.60) gives

$$dA_2 = -\, ds\,\frac{\rho}{r}\tan B_2\sin A_2 + dB_1\,\frac{\sin\ell}{\cos B_2} + dA_1\,\frac{\cos B_1}{\cos B_2} \tag{3.66}$$

The *inverse problem*, computation of s, A_1 and A_2 when ℓ, B_1 and B_2 are given leads, in a similar way, to differential formulas

$$\frac{\rho}{r}\, ds = -\, dB_1 \cos A_1 - dB_2 \cos A_2 + d\ell \cos B_1 \sin A_1 \qquad (3.67)$$

$$\frac{s}{r}\, dA_1 = dB_1 \sin A_1 + dB_2 \sin A_2 - d\ell \cos B_2 \cos A_2 \qquad (3.68)$$

$$\frac{s}{r}\, dA_2 = dB_1 \sin A_1 + dB_2 \sin A_2 + d\ell \cos B_1 \cos A_1 \qquad (3.69)$$

These formulas give slightly better results when the constant radius r is replaced by appropriate values of the radius of the curvature of the reference ellipsoid.

8. A straight line is staked using legs 25 meters long and setting each stake in the line of the two previous stakes. Suppose that at each stake, an accidental error of ± 1 cm is made in the transverse direction. Compute the accumulated error to be exspected at the end of a line 1 km long.

Ansver: $m^2 = 1^2 + 2^2 + 3^2 + \ldots + n^2 = \frac{1}{6}n(n+1)(2n+1)$

With $n = 40$, $m = \pm$ 1.5 meters. When n is a great number, $m = \pm\sqrt{\frac{n^3}{3}}$

9. An observer in Honolulu ($\phi = 21\overset{\circ}{.}3$ N, $\lambda = 157\overset{\circ}{.}9$ W, $h = 0$) was watching a passive satellite Febr. 8, 1968. It disappeared into the shadow of the Earth at $7^h30^m \pm 1^m$ P.M. local official time. The coelestial coordinates of the spot of disappearance were estimated to be: Right Ascension 112°, Declination $\delta = -7°$. The height H of the satellite above the earth's surface (supposed to be spherical with $R = 6378$ km) was computed. Estimate the mean error of H.

Denoting by t the Greenwich hour angle and by the prime the coordinates of the sun, following formulas can be used:

$$A = \sin\phi \sin\delta' + \cos\phi \cos\delta' \cos t'$$
$$B = \sin\phi \sin\delta + \cos\phi \cos\delta \cos t$$
$$C = \sin\delta \sin\delta' + \cos\delta \cos\delta' \cos(t' - t)$$

$$(1 - C^2)\ x^2 + 2\ (B - AC)\ x = A^2$$

$$H = R(\sqrt{1 + 2Bx + x^2} - 1)$$

Answer: $H = 363 \pm 10$ km

10. The geodetic operations are often divided into classes according to the accuracy to be pursued. There are "first order", "second order" etc. triangulations, in each country or city, similarly "first order" and "second order" traverses, different nets of leveling a.s.o. On the other hand, there is no general scale to be used when comparing different methods of operation, or operations carried out in different countries or at different times.

As an attempt to establish such a scale, we introduce here *class number* M in such a way that

$$M = 12{,}5 + \log(m_s/s)^2 \tag{3.70}$$

where s is either the quantity itself the mean error m_s of which has been estimated or, in most geodetic applications, the radius of region within which m_s means the mean error of the determined location. For angle measurements, $s = \rho$. In other words, m_s/s is either the relative mean error or the angular mean error in radians.

The limits for a first order triangulation would then be

1.00	to	3.16	parts in million in coordinates
0.21	to	0.65	sexagesimal seconds of arc
0.64	to	2.01	centesimal seconds of arc

The limits of other classes can be found from these values just moving the decimal points.

The base line measurements and the precise levellings very often belong, according to this convention, to "zero order" operations. At the opposite end of the scale we may have the aiming of the heavy artillery. *E.g.*, 10 m/10 km corresponds to $M = 6.5$.

4 ESTIMATION OF STANDARD ERROR

4.1 *Standard error of the arithmetic mean*

The approximate formula of the variance

$$\sigma^2 = \frac{[vv]}{n} \tag{4.01}$$

is not suitable for the practical estimation of the root mean square error σ if n is not very great, because the arithmetic mean

$$x_m = \frac{1}{n}(x_1 + x_2 + \ldots + x_n) \tag{4.02}$$

can deviate considerably form the theoretical mean

$$x_t = \lim_{n\to\infty} x_m \tag{4.03}$$

when n is a small number. The residuals

$$v_i = x_m - x_i \qquad (i = 1,2,\ldots n) \tag{4.04}$$

contain, in such a case, a systematic error. An unbiased estimate can be acquired by the aid of the theoretical errors

$$\varepsilon_i = x_i - x_t \tag{4.05}$$

in the following way. We write

$$v_i = x_m - x_t - \varepsilon_i = \frac{[\varepsilon]}{n} - \varepsilon_i$$

or, when $i = 1$,

$$v_1 = \frac{1}{n}\{(1-n)\varepsilon_1 + \varepsilon_2 + \ldots + \varepsilon_n\} \tag{4.06}$$

Now we can replace formula (4.01) by a better approximation

$$m^2 = \frac{[\varepsilon\varepsilon]}{n} \tag{4.07}$$

For the square of the first residual we obtain from (4.06)

$$\begin{aligned} v_1^2 = \frac{1}{n^2}\{&(n-1)^2\varepsilon_1^2 + \varepsilon_2^2 + \ldots + \varepsilon_n^2 \\ &+ 2(1-n)\varepsilon_1\varepsilon_2 + \ldots + 2(1-n)\varepsilon_1\varepsilon_n \\ &+ 2\varepsilon_2\varepsilon_3 + \ldots + 2\varepsilon_{n-1}\varepsilon_n\} \end{aligned} \tag{4.08}$$

When the index 1 is replaced by $2,3,\ldots n$ and the squares of all residuals summed up, then each square term $\varepsilon_i{}^2$ has the same coefficient

$$\frac{1}{n^2}\{(n-1)^2 + n - 1\} = \frac{n-1}{n}$$

and each product term $\varepsilon_i\varepsilon_j$ has the same coefficient $-\frac{2}{n}$. Hence

$$[vv] = \frac{n-1}{n}[\varepsilon_i\varepsilon_i] - \frac{2}{n}[\varepsilon_i\varepsilon_j] \qquad \text{where } i \neq j \tag{4.09}$$

If the theoretical errors ε are independent of each other, the value of $[\varepsilon_i\varepsilon_j]$ can be positive of negative with the same probability. Therefore we write, as an estimate or an "expected value", $[\varepsilon_i\varepsilon_j] = 0$, and (4.09) becomes

$$[vv] = \frac{n-1}{n}[\varepsilon\varepsilon] \tag{4.10}$$

According to (4.07), the unbiased estimate of the standard error is then

$$m = \sqrt{\frac{[vv]}{n-1}} \tag{4.11}$$

For this fundamental formula, another argumentation might be desirable. When a quantity x has been measured twice, the difference d can be considered as a stochastic variable with the variance

$$m_d{}^2 = 2m_x{}^2 \tag{4.12}$$

The theoretical mean of $d = 0$. If we have N differences, the variance can then be estimated by $m_d{}^2 = \frac{[dd]}{N}$. When x has been measured n times, we form the differences $d = v_i - v_j$ in all combinations; then we have $N = \frac{1}{2}n(n-1)$. Certainly, all these differences are not independent of each other, but now we can sure that the square sum

$$[dd] = (n-1)[vv] - 2[v_iv_j] \tag{4.13}$$

depends evenly on all measurements. Because $[v] = 0$, we have also

$$[v]^2 = [vv] + 2[v_iv_j] = 0 \tag{4.14}$$

Therefore, (4.13) gives

$$[dd] = n[vv] \tag{4.15}$$

or

$$m_d^2 = \frac{n}{N}[vv] = \frac{2}{n-1}[vv] \tag{4.16}$$

Comparison with (4.12) gives again

$$m = \sqrt{\frac{[vv]}{n-1}} \tag{4.11}$$

The standard error of the arithmetic mean can now be computed by the aid of formula (3.18):

$$M = \sqrt{\frac{[vv]}{n(n-1)}} \tag{4.17}$$

During the practical computation of the arithmetic mean, the common first figures of the observed values x_i can be omitted and later added back to x_m. In a similar way, any approximate mean value x_o can subtracted from the measured values and added back to the computed mean value.

The arithmetic mean can be checked easily after the computation of residuals $v_i = x_m - x_i$. First, all positive values of v_i are summed up. Denoting this sum by s, the sum of all negative values of v_i must be $-s$. Because of the rounding-off errors, a deviation smaller than $n/2$ in the last decimal place is possible.

The square sum $[vv]$ can be checked in several ways:

$$[vv] = [(x_i - x_o)^2] - n(x_m - x_o)^2 \tag{4.18a}$$
$$= [(x_i - x_o)^2] - \frac{1}{n}[x_i - x_o]^2 \tag{4.18b}$$
$$= [(x_i - x_o)^2] - (x_m - x_o)[x_i - x_o] \tag{4.18c}$$

If the residuals v show a normal distribution, the standard error of one measurement could be estimated also by *Peters' formula*

$$m = \frac{2.5 \cdot s}{\sqrt{n(n-1)}} \tag{4.19}$$

Because this formula is seldom used in practice we omit it's theoretical argumentation.

4.2 *Double measurements*

Formulas (4.11) and (4.17) still can be misleading when n is small. For geodetic purposes, one quantity often is measured only twice but there might be many different quantities – *e.g.* the angles of a traverse – the standard error of which can be expected to be constant. If the difference of each pair of measurements is denoted by d, the standard error of one quantity is, according to (4.11) where $|v| = \frac{|d|}{2}$, $n = 2$:

$$m_i^2 = \frac{d_i^2}{2} \tag{4.20}$$

If there are N quantities, each measured twice, we can take as a common standard error of one measurement

$$m = \sqrt{\frac{[dd]}{2N}} \tag{4.21}$$

and as a common standard error of each arithmetical mean

$$M = \frac{1}{2}\sqrt{\frac{[dd]}{N}} \tag{4.22}$$

As a numerical example, we take a traverse with 60 angles which had been measured two times each using only even numbers for the readings of the (centesimal) second of arc. The frequencies f of the different values of d were

d	0"	2	4	6	8	10	12	14	16	18	20	22	24	26	28	30	34	38	54
f	4	7	7	5	5	5	6	6	3	2	1	2	2	0	1	1	1	1	1

From the square sum $[fdd] = 14316$ we compute $m_d^2 = [fdd]\,/\,[f]$ 238.6 or $m_d = \pm\ 15''.4$. The standard error of one angle when measured twice is, according to (4.22), $M = \pm\ 7''.7$. According to the formulas (2.12) and (2.11), we compute

r = the middle difference = 10", from which m_d = 15"
t = the average difference = 11".6, from which m_d = 14".5

The greatest difference 54" = $3.5 \cdot m_d$ is probably due to a blunder (or carelessness), cfr. Chapter 11.3.

4.3 *Station adjustment with complete sets*

A palmate set of directions from a station of triangulation is usually observed by the *series method*. One series regularly consists of a double reading of each direction, *viz.* reading of the circle at all directions once from left to right, then in the reverse position of the theodolite from right to left. The purpose of this method is to eliminate certain errors of the instrument from the mean of both readings; the mean then can be taken as one independent observation. The mean readings are written in form of Table 4.1 with one row for each series and with one column for each bearing. Ordinarily, from each reading of a row is substracted that of the first bearing; the first column of the table then contains mere zeroes. The differences thus obtained are denoted by ℓ in the following formulas. Let the number of bearings (columns) be s and that of the series (rows) r. The table is expanded by the column of row sums R and by the row of column sums S; another additional row shows the mean values S/r.

Then we prepare Table 4.2 showing the differences $d = S/r - \ell$; the first column contains mere zeroes in this table too. The row sums of this table are denoted by R', the column sums are = 0. Finally, we prepare Table 4.3 showing the differences $v = d - R'/s$. In this table, both the row sums and column sums must be = 0.

The square sum of the residuals v can be checked by several formulas:

$$[vv] = [dd] - \frac{[R'R']}{s} \qquad (4.23a)$$

$$[dd] = [\ell\ell] - \frac{[SS]}{r} \qquad (4.23b)$$

$$[R'R'] = [RR] - \frac{[R]^2}{r} \qquad (4.23c)$$

The standard error of one reading ℓ is, as shown below,

$$m = \sqrt{\frac{[vv]}{(r-1)(s-1)}} \qquad (4.24)$$

The final outcome of the station adjustment consists of the mean bearing S/r on the last row of Table 4.1 and of the common standard error

$$M = \frac{m}{\sqrt{r}} \qquad (4.25)$$

of each of them. There are no correlations between these adjusted values of the observed directions.

As to the derivation of formula (4.24), we imagine that for each row of Table 4.3 the differences of two v-values have been computed in all combinations the number of which is $\frac{1}{2}s(s-1)$. Then the differences w of these differences should again be computed in all combinations, this time for each column; the number of the second differences is $\frac{1}{2}r(r-1)$ for each column or $\frac{1}{4}rs(r-1)(s-1)$ all together. The theoretical mean value of the differences w is $= 0$; in other words, we can estimate the variance of w by formula (4.07) or

$$m_w^{\ 2} = \frac{4[ww]}{rs(r-1)(s-1)}$$

On the other hand, by the aid of formula (4.15) we could show that $[ww] = rs[vv]$. Because each value of w is a double difference of four values of v and because there obviously are no systematical correlations between the w-values, we have $m_w = 2m_v = 2m$. In this way, we eventually obtain formula (4.24).

In the station adjustment, there appear as unknown parameters the r zero-corrections for the different series (rows) and the $s-1$ angles (or non-zero values of S/r) between the bearings. The number of observations is rs, hence the number of independent misclosures is $rs - r - (s-1) = (r-1)(s-1)$ which appears as the denominator in formula (4.24). In fact, formula (4.24) is a special case from the general rule: *if the measurements have equal weights, the variance of one measured value can be estimated by the quotient: sum of the squares of residuals divided by number of the extra measurements* (or misclosures). The argumentation of this rule will be given in Chapter 13.4.

As an example, we take $r = 6$ series of readings for $s = 4$ bearings at one point of a city triangulation.

Table 4.1, readings ℓ

Bearing	A	B	C	D	R
Series					
1	0°00'00"	19°41'48".8	37°49'36".7	75°03'30".2	15".7
2	00	47.2	39.7	31.7	18.6
3	00	47.2	37.2	29.7	14.1
4	00	45.1	40.2	32.8	18.1
5	00	46.3	37.3	27.2	11.3
6	00	49.5	37.5	28.2	15.2
S	00	44.1	48.6	00.3	93.0
Mean	0°00'00".00	19°41'47".35	37°49'38".10	75°03'30".05	
	$[\ell\ell] - [RR]/4 - [SS]/6 + [R]^2/24 = 34.025$				

Table 4.2, differences d

Bearing	A	B	C	D	R'
Series					
1	00	-1.45	+1.40	-0.15	-0.20
2	00	+0.15	-1.60	-1.65	-3.10
3	00	+0.15	+0.90	+0.35	+1.40
4	00	+2.25	-2.10	-2.75	-2.60
5	00	+1.05	+0.80	+2.35	+4.20
6	00	-2.15	+0.60	+1.85	+0.30
	$[dd] - [R'R']/4 = 34.025$				

Table 4.3, residuals v

Bearing	A	B	C	D
Series				
1	+0.05	-1.40	+1.45	-0.10
2	+0.78	+0.92	-0.82	-0.88
3	-0.35	-0.20	+0.55	0.00
4	+0.65	+0.90	-1.45	-2.10
5	-1.05	0.00	-0.25	+1.30
6	-0.08	-2.22	+0.52	+1.78

$[vv] = 34.025$ $(r-1)(s-1) = 15$

$m^2 = 2.268$ $m = \pm 1''.50$

$M^2 = 0.378$ $M = \pm 0''.61$

4.4 *Ferrero's formula*

The misclosures can be considered as true or theoretical errors because the theoretical mean value of each misclosure is supposed to be zero. In fact, they are not quantities which are measured immediately but mathematical functions of immediately measured quantities. The variances of the misclosures can therefore be computed if the variance of the measurements are known. The inverse problem, computation of the variance of the measured quantities from all misclosures will be discussed in Chapter 13.4. Here we shall take a special case in which we can estimate approximately the variance of the observed angles (or bearings) before the complete adjustment of the triangulation net.

When all three angles of a triangle have been measured, the misclosure w = the sum of angles minus the theoretical value (180° plus the spherical excess) can be computed. From n independent triangles we obtain n values of w and by formula (4.07) we can estimate the variance of the misclosures

$$m_w^{\ 2} = \frac{[ww]}{n} \tag{4.26}$$

On the other hand, each misclosure depends on the sum of three angles and each angle is the difference of two observed bearings. Hence we have, according to formula (3.17)

$$m_w = \sqrt{6}\, m \tag{4.27}$$

where m is the standard error of one bearing. In this way we obtain Ferrero's formula

$$m = \sqrt{\frac{[ww]}{6n}} \tag{4.28}$$

This formula is not quite correct theoretically, because some bearings affect two or more triangles, some bearings only one. In addition, this estimate does not use the information which is enclosed in the misclosures of the side equations or other similar conditions of the net.

A numerical example is given in Chapter 11.4.

5 *WEIGHTS*

5.1 *Definitions*

An elementary application of the weights and the weighted mean is illustrated by the following example. An angle L was measured n_a times in one night and n_b times in the next night. Each night, the arithmetic mean was taken as a provisional result:

$$L_a = \frac{[L]_a}{n_a} \qquad \text{and} \qquad L_b = \frac{[L]_b}{n_b} \tag{5.01}$$

The definitive result can now be computed in three different ways. First, we can compute the arithmetic mean of all original measurements. Secondly, we can take the weighted mean of the provisional results, using the numbers n as weights. Both these computations give the same result:

$$L_m = \frac{[L]}{n_a+n_b} = \frac{n_a L_a + n_b L_b}{n_a + n_b} \tag{5.02}$$

In the latter computation, the numbers n can be replaced by any *weights* p which satisfy the condition

$$\frac{p_a}{n_a} = \frac{p_b}{n_b} = \text{constant} \tag{5.03}$$

In this computation, the *weights are arbitrary numbers which are proportional to the numbers of original, equally weighted measurements.* Obviously, the number of provisional or group results can be greater than two. The *general formula of the weighted mean* then reads

$$L_p = \frac{[p_i L_i]}{[p_i]} \tag{5.04}$$

If the original measurements have equal weights, they have the same standard error m. According to formula (3.18) the provisional mean values then have standard errors

$$M_a = \frac{m}{\sqrt{n_a}} \quad \text{and} \quad M_b = \frac{m}{\sqrt{n_b}} \tag{5.05}$$

This gives

$$\frac{M_a}{M_b} = \frac{\sqrt{n_b}}{\sqrt{n_a}} \tag{5.06}$$

or combined with (5.03),

$$\frac{p_a}{p_b} = \frac{M_b^{\,2}}{M_a^{\,2}} \tag{5.07}$$

According to this formula, *the weights are arbitrary quantities which are inversely proportional to the variances.* This relation is now taken as the *general definition of the weights.* Later, we shall let this definition be valid also in the case that the quantities with different weights are correlated.

Of course, the weights – or the variances – must be estimated before the computation of the weighted mean or, generally, before any adjustment. The estimation of the weights depends largely on the judgement of the person who undertakes the responsibility of the computation. In order to illustrate the ambiguity which can arise in this connection, we supply our example with following numerical values (omitting the degrees and minutes of the angles):

	1. night	2. night	Sum
L	30".2 31.7 29.5 32.7 27.6 28.2 29.0 31.1	31".6 30.1 30.2 29.3	
$[L]_i$	240".0	121".2	361".2
n_i	8	4	12
L_i	30".0	30".3	30".1
p_i	2	1	3
$p_i L_i$	60".0	30".3	90".3

Using formulas (4.11) and (4.17), we estimate the variances for each night separately:

	1. night		2. night	
	v	vv	v	vv
	+0.2	0.04	+1.3	1.69
	+1.7	2.89	-0.2	0.04
	-0.5	0.25	-0.1	0.01
	+2.7	7.29	-1.0	1.00
	-2.4	5.76		
	-1.8	3.24		
	-1.0	1.00		
	+1.1	1.21		
Sum	0	21.68	0	2.74
m^2		3.10		0.91
M^2		0.39		0.23

Now we must make an important decision. If to the provisional mean of the second night is given weight $p = 1$, to that of the first night should perhaps be given weight $p = 0.23/0.39 = 0.6$, according to the general definition of weights, and not $p = 2$ as we had in the first table. Therefore, we must consider the possibility that the individual measurements did not have the same weight in the both nights. Because the number of measurements was not very great, the estimates given in the second table are not very reliable. The observer, however, may have made some remarks upon the weather or other circumstances which make it probable that the weights really were different.

On the other hand, we see that the different estimates of the weights have only a trifling significance upon the results, for if the weights of the latter estimate are accepted, the definitive mean value will be 30.2 instead of 30.1, the difference being a small fraction of the standard errors $M_1 = \pm 0''.6$ and $M_2 = \pm 0''.5$.

5.2 *Geodetic applications*

In Chapter 3.2, we have mentioned that in *tape measurements* and in *spirit levelling* the standard error usually is supposed to be proportional to the square root of the measured distance. The weights, therefore, are inversely proportional to the distances themselves. The same rule is often applied to the adjustment of coordinates at the intersection point of traverses.

In the *measurement of the horizontal angles*, the standard error of one bearing is considered as independent of the distance of the target, provided that the distance is long enough to make the possible error of the centering insignificant. In the adjustment of the triangulation nets, therefore, all bearings have equal weights, usually $p = 1$, if the number of series (cfr. Chapter 4.3) is the same at all stations. Of course, the external circumstances are supposed to be same too.

The *vertical angles* can be dealt with in the same way when the distances do not exceed a few kilometers. In the adjustment of *trigonometrical levelling*, however, it is more usual to take the height differences as measured quantities, not the vertical angles. In this case, the weight of one height difference is inversely proportional to the square of the distance. When the distances are about 10 km or more, the uncertain correction for the refraction causes that the weight decreases still more rapidly.

As to the *optical and electronic determinations of distances*, special experiments or theories are required for the determination of the weights or variances as functions of the distance. *E.g.*, in Finland following formulas have been accepted:

$$\text{Tellurometer n:r 2:} \quad m_t^2 = 25 + 0.16\, s^2 \tag{5.08}$$

$$\text{Geodimeter n:r 4:} \quad m_g^2 = 2.25 + 0.25\, s^2 \tag{5.09}$$

In these formulas, s must be expressed in kilometers, then m_g and m_t are obtained in centimeters. If we suppose that the weight $p = 1$ is given to one bearing the standard error of which has been estimated (*e.g.* by Ferrero's formula) to be m, the weight of a tellurometer distance can be computed by formula

$$p_t = \frac{m^2}{m_t^2} \tag{5.10}$$

It is important to note the *dimension of the weight.* If m is given in sexagesimal seconds and m_t in decimeters, the dimension of the weight is $(1''/\mathrm{dm})^2$. *E.g.*, with $m = 1\overset{''}{.}2$ we obtain for different distances s:

Table 5.1 *Tellurometer n:r 2*

s	m_t^2	p_t	m_t	ms/ρ''
1	0.2516	5.7	0.50	0.06
2	0.2564	5.6	0.51	0.12
5	0.29	5.0	0.54	0.29
10	0.41	3.5	0.64	0.58
15	0.61	2.4	0.78	0.87
20	0.89	1.6	0.94	1.16
25	1.25	1.15	1.12	1.45
30	1.69	0.85	1.30	1.74
40	2.81	0.51	1.68	2.33
km	dm^2	$(\frac{1''}{\mathrm{dm}})^2$	dm	dm

Table 5.2 *Geodimeter n:r 4*

s	m_g^2	p_g	m_g	ms/ρ''
1	0.0250	57.6	0.16	0.06
2	0.0350	41.1	0.19	0.12
5	0.0850	16.9	0.29	0.29
10	0.2725	5.3	0.52	0.58
15	0.5825	2.5	0.76	0.87
20	1.0225	1.4	1.01	1.16
25	1.5825	0.91	1.26	1.45
30	2.2725	0.63	1.51	1.74
40	4.0225	0.36	2.01	2.33
km	dm^2	$(\frac{1''}{\mathrm{dm}})^2$	dm	dm

The columns m_t and m_g of these tables show the longitudinal error, the column ms/ρ'' the lateral error at a point for which both the distance and the bearing have been measured. The former table shows that for the shorter distances the longitudinal error is essentially larger than the lateral one. We can conclude that for triangulations in which the sides are a few kilometers long only, the measurements with a tellurometer are almost futile if the accuracy of the theodolite bearings is as good as 1 to 2 sexagesimal seconds. Note that the weights p_t or p_g can not used for similar conclusions because the numerical value of the weight depends on the units used for the standard errors m, m_t and m_g.

5.3 *Standard error of the weighted mean*

The unit weight $p = 1$ can be given to some real or to a fictitious observation. If there is a majority of observations which have equal weights, it is advisable to give the unit weight to those observations, *e.g.* in the triangulation, to those bearings which have been observed with a standardized number of series. In spirit levelling and in the adjustment of traverses, unit weight is often given to height differences or to coordinates which have been measured and computed for distances of one kilometer. If the distance is s kilometers, the weight is

$$p = \frac{1}{s} \tag{5.11}$$

If all measured quantities which will be included in the same adjustment have the same dimension, the weights can be taken as dimensionless numbers. If the quantities have different dimensions, *e.g.* if both angles and distances have been measured in a triangulation net, the products pvv must have the same dimension. Therefore, the weights generally have different dimensions; *e.g.*, each p has the dimension of the corresponding measured quantity with the exponent -2, and the products pvv are dimensionless.

In the following formulas, the symbol m without any subscript means the *standard error of unit weight*. Usually, the numerical value of m is unknown but the relative values of the standard errors of all measured quantities must be estimated before any adjustment. After the adjustment, m can be estimated and then the absolute values of all standard errors can be computed by formula

$$m_p = \frac{m}{\sqrt{p}}. \tag{5.12}$$

Especially, the standard error of the weighted mean is

$$M_p = \frac{m}{\sqrt{[p]}} \tag{5.13}$$

for the weight of the weighted mean is $[p]$. The numerical value of m can be estimated as follows.

Let the measured values be L_i $(i = 1,2,\ldots n)$ and the respective weights p_i. Then the weighted mean is

$$L_p = \frac{[pL]}{[p]} \tag{5.04}$$

and the residuals

$$v_i = L_p - L_i \tag{5.14}$$

As a check, we have

$$[pv] = 0 \tag{5.15}$$

Then we can compute the weighted square sum $[pv \cdot v]$ checking it by formulas

$$[pvv] = [pLL] - [p]L_p^2 \tag{5.16a}$$

$$= [pLL] - [pL]L_p \tag{5.16b}$$

Of course, the common first figures of L_i are omitted during these computations.

Assume that each L_i is now multiplied by the respective square root $\sqrt{p_i}$. Then we have n fictitious observations which all have weight 1. If the true error of an actual measurement is ε_i and the true error of L_p is x, then the true error of the fictitious observation is

$$\sqrt{p_i}\,\varepsilon_i = -\sqrt{p_i}\,(v_i + x) \tag{5.17}$$

The square sum of these expressions is

$$[p\varepsilon\varepsilon] = [pvv] + 2x[pv] + [p]x^2 \tag{5.18}$$

Substituting here $[pv] = 0$ and estimating x by M_p according to (5.13) we obtain $[p\varepsilon\varepsilon] = [pvv] + m^2$. According to formula (4.07), we can estimate m by

$$m^2 = \frac{[p\varepsilon\varepsilon]}{n} \tag{5.19}$$

In this way we obtain the equation

$$nm^2 = [pvv] + m^2 \tag{5.20}$$

from which follows

$$m = \sqrt{\frac{[pvv]}{n-1}} \tag{5.21}$$

This is the standard error of unit weight. The standard error of the weighted mean is, according to (5.13)

$$M_p = \sqrt{\frac{[pvv]}{[p](n-1)}} \tag{5.22}$$

5.4 *Examples*

1. The angles of a triangulation net have usually been measured with standard theodolite which gives the standard error of ± 4" for one reading. At some stations, a smaller theodolite had to be used for which the corresponding standard error was estimated to be ± 6". The standard number of series was 6. What should be taken as the number of series for the smaller theodolite in order to make the mean of the readings equally weighted?

According to (5.06), we obtain the equation

$$\frac{n}{6} = \left(\frac{6}{4}\right)^2$$

from which $n = \frac{27}{2}$ or, as an even integer, $n = 14$.

2. The same distance was measured with a pair of rods, with a tape and with an optical rangefinder. The standard error of each measurement was estimated on the basis of the old experience. The results:

Pair of rods	78.294 meters	± 3 mm
Tape	78.301	4
Rangefinder	78.288	5

Compute the weighted mean and estimate it's standard error.

If the third measurement is taken as the weight unit, the weights are $\frac{25}{9} = 2.8$, $\frac{25}{16} = 1.6$ and 1.0; sum $[p] = 5.4$. Using an approximate mean 78.290 we obtain $2.8 \cdot 4 + 1.6 \cdot 11 - 1.0 \cdot 2 = 26.8$ which divided by 5.4 gives 5.0 or the weighted mean 78.295 meters. The residuals are +1, -6 and +7 mm from which $[pvv] = 109$. According to (5.21), $m^2 = 55$, but this estimate is based on two extra measurements only. The estimate based on the old experience gives for the weight unit $m^2 = 25$. As the latter can here be considered as more reliable, we use it in formula (5.12) obtaining $M_p{}^2 = \frac{25}{5.4} = 4.6$ or $M_p = \pm 2$ mm.

3. In a triangle ABC, the angles A and B have been measured with equal weights $p = 1$. Then the angle C can be computed and the weight of the computed value is $\frac{1}{2}$ because the variance is 2 times the variance of one measured angle. If now the angle C is measured too with weight 1, there appears a misclosure which must be adjusted or divided evenly to all three angles. The weight of each adjusted angle is then 1.5. The algebraic proof is left to the reader.

4. The length of the total line of spirit levelling is S. At each end of the line, there is a bench mark the height of which is supposed to be correct (the points may be identical in which case the line forms a loop). At distance s from one end there is a new bench mark. Which weight can be given to the height of the new bench mark if the unit weight is given to the spirit levelling of a line 1 kilometer long?

The height of the new point is obtained in two ways with weights $1/s$ and $1/(S-s)$. The weight of the mean is

$$p = \frac{1}{s} + \frac{1}{S-s} = \frac{S}{s(S-s)}$$

This weight has it's minimum value in the halfway $s = \frac{S}{2}$ where $p = \frac{4}{S}$.

As a numerical example, we take a loop containing the initial point P and three new points, X, Y, and Z. The levelling was carried out twice with results given in the following table.

Section	s/km	Lev.1.	Lev.2.	Mean	Adjustment		v_1	v_2	pv_1^2	pv_2^2
PX	1.72	+1.546	+1.558	+1.552	-4	+1.548	+2	-10	2	58
XY	2.83	+1.348	+1.352	+1.350	-6	+1.344	-4	- 8	6	23
YZ	2.08	+6.225	+6.235	+6.230	-4	+6.226	+1	- 9	0	39
ZP	3.26	-9.112	-9.110	-9.111	-7	-9.118	-6	- 8	11	20
Sum	9.89	+7	+35	+21	-21	0	-7	-35	[pvv]= 159	

The misclosure of the mean levelling is +21 mm. The adjustment is carried out dividing the misclosure proportionally with respect to s. The residuals have been computed separately for each levelling; their weighted square sum is 159. The standard error of unit weight is $m = \sqrt{159/5} = \pm 5.6$ mm/$\sqrt{\text{km}}$, for there are 8 measured height differences and 3 unknown heights which make the number of extra measurements = 5. In the table below, we still give the weights of the heights of the new points and the height differences from P with the standard errors:

$p_X = 1.41$,	PX:	+1.548 meters	± 4.8 mm
$p_Y = 0.81$,	PY:	+2.892	6.3
$p_Z = 0.91$,	PZ:	+9.118	5.9

5.5 *Measurement of angles in all combinations*

A palmate set of s bearings can be measured, (instead of using the series method described in Chapter 4.3), in such a way that the individual angles are measured in all combinations, the number of which is $n = \frac{1}{2}s(s-1)$. We take immediately a numerical example with $s = 4$, $n = 6$:

			Adjustment: v =	vv =
Angle	APB =	25°17'38"	-4"	16
	APC =	164 30 25	-1	1
	BPC =	139 12 52	-2	4
	CPD =	84 59 11	-3	9
	DPA =	110 30 33	-5	25
	DPB =	135 48 00	+2	4
			[vv] =	59

Each angle (*e.g.* APB) can now be obtained in s-1 different ways which are completely independent of each other. In addition to that value which has been measured immediately ($25^\circ 17'38''$), it can be computed as the difference or sum of two other angles (APC - BPC = $25^\circ 17'\ 33''$ and DPB - DPA = $25^\circ 17'\ 27''$). If each measured angle has the weight $p = 1$, the difference or sum has the weight $p = \frac{1}{2}$. The adjusted value is the weighted mean of all these values ($25^\circ 17'34''$) and the weight of this mean is $\frac{s}{2}$ (= 2). In the table above, we have shown only the corrections v which must be given to the measured values. Usually, the results are given in form of a series of bearings:

PA	0°00'00"
PB	25 17 34
PC	164 30 24
PD	249 29 32

where the weight of each bearing is s (= 4). It can be shown that these fictitious observations of bearings are independent of each other (all covariances = 0).

These weights are equivalent to those which would have been obtained by the series method when the number of series is $\frac{s}{2}$. In the method of all combinations the number of necessary (double) readings is $2n = s(s-1)$, or 12 in our example, but in the series method with $\frac{s}{2}$ series the corresponding number would be only $\frac{1}{2}s^2$ or 8 in our example. Therefore, the method of all combinations is considered as a waste of labor except in cases when special circumstances (*e.g.* a weak observation stand) make the measurement of single angles unavoidable.

The number of extra measurements in this method is $r = n - (s-1) = \frac{1}{2}(s-1)(s-2)$, or 3 in our example. The variance of one measurement is therefore

$$m^2 = \frac{[vv]}{r} = \frac{2[vv]}{(s-1)(s-2)} \qquad (5.23)$$

or 19.7 in our example. The standard error of one fictitious bearing is

$$m_f = \frac{m}{\sqrt{s}} \qquad (5.24)$$

or ± 2"2 in our example.

If the method of all combinations is used in a triangulation net, some precautions are required in order to make all fictitious bearings equally weighted. Let K be the number of series at the stations where the series method is used. Then at each station where there are s bearings to be observed in all combinations, each angle must be measured k times with

$$k = \frac{2K}{s} \tag{5.25}$$

If $K = 12$, we have for k the following table

s	2	3	4	5	6	7	8
k	12	8	6	5	4	4 or 3	3

6 ADJUSTMENT BY VARIATION OF THE PARAMETERS

6.1 *Derivation of the formulas*

In Chapter 1.1, we have mentioned a few elementary problems of adjustment dividing them into two groups A and B. Both groups need technically different methods of solution which, however, are based on the same principle of least squares. In this chapter, we shall derive general formulas for the solution of problems of type A. This solution is known by names "method of indirect observations" or "adjustment by variation of the parameters"; to be brief, we may call it also "method A".

Suppose that the measured quantities can be represented by a mathematical model or "theory" which contains some unknown parameters. In this chapter, we use only three parameters X, Y, Z, because the general construction of the formulas can be made obvious already with this restriction, but in Chapter 13.1 we repeat the derivation, using matrix notations, for an arbitrary number u of parameters.

On the basis of the mathematical model, we can express the measured quantities L_i $(i = 1,2,...n)$ as functions of the parameters:

$$L_i = L_i(X,Y,Z) \tag{6.01}$$

The measured numerical values of these quantities will be denoted by $L_{b,i}$. For each measurement, we must estimate the weight p_i as explained in Section 5.

If the number n of measurements is greater than the number u of the parameters, there generally is no such a group of numerical values of parameters which would exactly give values $L_{b,i}$ when inserted into equations (6.01). The principal problem of adjustment is to find such a group of values X_a, Y_a, Z_a of parameters that the residuals

$$v_i = L_i(X_a, Y_a, Z_a) - L_{b,i} = L_{a,i} - L_{b,i} \tag{6.02}$$

comply to the principle

$$[pvv] = \text{minimum} \tag{6.03}$$

If the functions (6.01) were linear or of form

$$L_i = a_i X + b_i Y + c_i Z + \lambda_i \tag{6.04}$$

where a, b, c and λ are contants, the solution would be easy enough. Therefore, we shall first linearize the equations (6.02) in the following way. We try to find good approximate values X_o, Y_o, Z_o for the parameters and take the variations

$$\begin{aligned} x &= X_a - X_o \\ y &= Y_a - Y_o \\ z &= Z_a - Z_o \end{aligned} \tag{6.05}$$

as temporary unknown quantities. For the measured quantities, we must compute the corresponding values

$$L_{o,i} = L_i(X_o, Y_o, Z_o) \tag{6.06}$$

In addition, we need the numerical values for the partial derivatives

$$a_i = \frac{\partial L_i}{\partial X}, \quad b_i = \frac{\partial L_i}{\partial Y}, \quad c_i = \frac{\partial L_i}{\partial Z} \tag{6.07}$$

using everywhere the approximate values X_o, Y_o, Z_o and $L_{o,i}$ in the numerical computations. The first terms of the Taylor expansion gives now linear equations

$$L_{o,i} + a_i x + b_i y + c_i z = L_{a,i} \tag{6.08}$$

Denoting

$$\ell_i = L_{o,i} - L_{b,i} \tag{6.09}$$

we obtain by the aid of (6.02)

$$a_i x + b_i y + c_i z + \ell_i = v_i \tag{6.10}$$

These equations are called in English *observation equations*, in

German Fehlergleichungen (error equations).

Substituting the expressions (6.10) into (6.03), we obtain

$$\begin{aligned}[pvv] = [paa]x^2 &+ 2[pab]xy + 2[pac]xz + 2[pa\ell]x \\ &+ [pbb]y^2 + 2[pbc]yz + 2[pb\ell]y \\ &+ [pcc]z^2 + 2[pc\ell]z \\ &+ [p\ell\ell] = \min.\end{aligned} \quad (6.11)$$

The minimum value is found by writing the partial derivatives of this expression with respect to x, y and $z = 0$. In this way we obtain equations

$$\begin{aligned}[paa]x + [pab]y + [pac]z + [pa\ell] &= 0 \\ [pab]x + [pbb]y + [pbc]z + [pb\ell] &= 0 \\ [pac]x + [pbc]y + [pcc]z + [pc\ell] &= 0\end{aligned} \quad (6.12)$$

which are called *normal equations*. In the general case, there are u linear equations whith u unknown quantities. As a special feature, we see that the coefficients $[pab]$, $[pac]$, $[pbc]$ appear each twice in places which are symmetrical with respect to the principal diagonal of coefficients $[paa]$, $[pbb]$, $[pcc]$. The technical methods for the solution of this type of linear equations will be discussed in Section 8.

When the solution x, y, z which is found from the normal equations is substituted into equations (6.10), the numerical values of residuals v can be found. It is customary to compute the weighted square sum $[pvv]$ too. It gives a valuable check because it can be computed in several different ways, *e.g.* by formula

$$[pvv] = [p\ell\ell] + [pa\ell]x + [pb\ell]y + [pc\ell]z \quad (6.13)$$

which can be derived easily substituting equations (6.12) into (6.11). Furthermore, from it we obtain the standard error of unit weight

$$m = \sqrt{\frac{[pvv]}{n - u}} \quad (6.14)$$

The derivation of this important formula is omitted here because it can be carried out much easier by the aid of matrix notations, cfr. Section 13.

6.2 *Instructions for practical computations*

The numerical computations must be written or printed in form of tables according to some systematical arrangement for it is important that another person can easily check or continue the computations. In the following instructions, the consecutive steps of the computation will be numbered and entitled as schemes containing the summaries upon the main results; the necessary auxiliary computations may be filed as annexes to the main schemes which already may require many sheets with a carefully planned subdivision.

1. The *sketch* is inevitable for the adjustment of geometric nets. It is often drawn on a transparent paper set on a map. Using conventional notations, the fixed points and figures must be distinguished from the new points and figures which will be altered by the adjustment. The symbols for points, sides, bearings, *etc.* must be at proper places in the sketch. The direction of the computation of differences, levellings, traverses, chains *etc.* must be shown by arrows.

2. *Fixed quantities*, coordinates, sides, angles, heights *etc.* are picked up from old catalogues and computations. The trigonometric functions of fixed angles and other constants can be listed in this table too.

3. *Observations*, or the measured quantities $L_{b,i}$, are preferably listed in order of a succession number i. All corrections which do not depend on the adjustment, *e.g.* those due to centering, map projection, temperature, refraction *etc.* must be inserted beforehand.

4. *Approximate values of parameters* require often very extensive auxiliary computations, *e.g.* solution of triangles, computation of coordinates (intersections or resections), solution of pyramid problems in photogrammetry, computation of orbits for the satellites *etc.* There might be several alternative ways for the selection of convenient parameters but usu-

ally they are obvious enough.

As to the computation of the approximate values of the selected parameters, there again are many different ways. The only important thing is that these values are sufficiently close to the final results of the adjustment; in the worst cases the adjustment gives only new approximate values which must be improved by a second adjustment, cfr. Tables VII and VIII in Chapter 17.3. The best way is to select the mininum number of observations and compute the parameters backwards as an inverse problem to formulas (6.01). This computation may be carried out by simplified and approximate methods and the results can be rounded off. The parameters are taken in such an order that the most important ones for which the variances and covariances should be computed are left to the last places.

5. *Approximate values of the measured quantities* must be computed by the exact formulas (6.06) maintaining extreme precision and using reliable checking devices. If the approximate values of parameters have been computed from u selected observations by exact methods, this computation gives those measured values back but the other observations reveal small misclosures. The coefficients a, b, c ... of the observation equations can often be computed in connexion of this main computation; for them, an accuracy of three or four significant figures is sufficient.

6. *The observation equations* are written in the form of a table. For each observation must be reserved a row, for each parameter a column. Only the coefficients are written in the table; this array forms the the matrix A as will be explained in Chapter 13.1. Most coefficients are = 0; these places are usually left empty. Often the significant coefficients are +1 or -1; these elements can be written in the right places by the aid of the sketch. The units of the parameters and the measured quantities should be chosen in such a way that all coefficients are of the order 1.

The differences ℓ computed by formula (6.09) are written in their own column; when proper units are selected these figures are of order 1 too. Then an effective sum control for the computation of the next tables can be obtained. For this purpose, we write into the next column the row sums

$$s_i = a_i + b_i + c_i + \ldots + \ell_i$$

If the observations have different weights, another column is necessary for the estimated values of p_i. In this case, the following arrangement is often useful for the computations with desk calculators. For each equation we reserve an additional row on which the figures of the original equation have been multiplied by p_i. As an alternative, each equation could be multiplied by $\sqrt{p_i}$ and the new equations written in a special scheme.

7. *Normal equations* form a triangular scheme because the coefficients below the principal diagonal can be found from the symmetrical places above the diagonal. If a slide rule only is available, extensive additional tables for the individual products paa, pab, ... pas are necessary. When a desk calculator is used, the products $pa \cdot a$, $pa \cdot b$, ... $pa \cdot s$ can be summed up immediately by the aid of accumulative multiplication. For this purpose, the double rows of the observation equations mentioned above are advisable; in the alternative method, the products $\sqrt{p}a \cdot \sqrt{p}a \ldots \sqrt{p}a \cdot \sqrt{p}s$ are summed up. Each row can be checked by the aid of the sum control column S showing the product sums $[pas]$, $[pbs]$ *etc.* An extra row shows the product sums $[p\ell\ell]$ and $[p\ell s]$.

8. *The solution of the normal equations* will be explained in Section 8. The results, numerical values of x, y, z ... are written into scheme 4 as a new column. Then the definitive values of the parameters can be written into the next column.

For checking purposes, the values of x, y, z... are substituted in the observation equations, scheme 6, and the resulting values of v and pv written into additional columns. Then a checking value for $[pvv]$ can be formed. A more complete check can be obtained by repeating the computations of scheme 5 with the new values of parameters but if an error in the old computation is found first now, the entire adjustment must be renewed.

7 ADJUSTMENT BY CORRELATES

7.1 *Derivation of formulas*

In group B of the problems mentioned in Chapter 1.1, attention was directed first to the *misclosures*, not to the parameters. In the mathematical model, anyway, there is a certain number u of independent parameters but they can be selected and computed afterwards if the misclosures are first got rid of by an adjustment. This special method of adjustment is known by names "method of conditioned observations" or "adjustment by correlates"; to be brief, we may call it also "method B".

To begin with, we need only the number u of the parameters. If there are n measured quantities the number of independent misclosures is

$$r = n - u \tag{7.01}$$

The first task is to find r mathematical equations between the measured quantities L_i ($i = 1,2,\ldots n$) without using any of the parameters. These equations should be as simple as possible and no equation should be algebraically deductible from the others. It is not necessary that each L_i appears in all equations but each L_i must appear in one equation at least. In many applications, *e.g.* in photogrammetry, it is difficult to find these equations without using the parameters; for those applications, this method B is not suitable.

We denote these equations

$$W_j(L_1, L_2,\ldots L_n) = 0 \qquad (j = a, b,\ldots r) \tag{7.02}$$

When the measured values $L_{b,i}$ are substituted into these equations, small misclosures $w_a, w_b \ldots w_r$ are revealed:

$$w_j = W_j(L_{b,1}, L_{b,2},\ldots L_{b,n}) \tag{7.03}$$

For each of the measured quantities, we must estimate the weight

p_i as explained in Section 5. The principal problem of the adjustment is to find such corrections v_i for the measured quantities $L_{b,i}$ that the adjusted values

$$L_{a,i} = L_{b,i} + v_i \tag{7.04}$$

satisfy the equations (7.02). Because the number of unknown quantities v_i is greater than the number of equations, there is an infinite number of possible solutions. From these solution, we shall pick that one which complies to the principle

$$[pvv] = \text{minimum} \tag{7.05}$$

If the equations (7.02) were linear or of form

$$W_a(L_1, L_2, \ldots L_n) = a_1 L_1 + a_2 L_2 + \ldots + a_n L_n + \phi_a \tag{7.06}$$

where the coefficients a and the last term ϕ_a are constant, the solution would be easy enough. Therefore, we shall first linearize the equations (7.03) in the following way. We compute the numerical values of the partial derivatives

$$a_i = \frac{\partial W_a}{\partial L_i}, \qquad b_i = \frac{\partial W_b}{\partial L_i}, \qquad c_i = \frac{\partial W_c}{\partial L_i}, \qquad \ldots \tag{7.07}$$

for the measured values $L_{b,i}$. If now the measured quantities in equations (7.03) are replaced by the adjusted values according to (7.04), the misclosures should disappear. Provided that the misclosures are not very large, this leads to linear equations

$$\begin{aligned} a_1 v_1 + a_2 v_2 + \ldots + a_n v_n + w_a &= 0 \\ b_1 v_1 + b_2 v_2 + \ldots + b_n v_n + w_b &= 0 \\ c_1 v_1 + c_2 v_2 + \ldots + c_n v_n + w_c &= 0 \end{aligned} \tag{7.08}$$

which are called *condition equations*. The number of condition equations is r but here we shall use only these three equations because the general construction of the formulas can be made obvious already with this restriction.

We denote the expressions on the left hand side of equations (7.08) by K_a, K_b, K_c. When the equations (7.08) are satisfied, $[pvv]$ has the minimum value for the same group of solutions v_i

as the expression

$$S = \frac{1}{2}[pvv] - k_a K_a - k_b K_b - k_c K_c \tag{7.09}$$

The unknown constants k_j are called *correlates* and they will be discussed later. The expression S which is a function of the residuals v_i has the minimum value when all the partial derivatives with respect to the residuals disappear. In this way, we obtain n equations

$$p_i v_i = a_i k_a + b_i k_b + c_i k_c \tag{7.10}$$

or

$$v_i = \frac{a_i}{p_i} k_a + \frac{b_i}{p_i} k_b + \frac{c_i}{p_i} k_c \tag{7.11}$$

The expressions (7.11) are called *developments by correlates* for as soon as we have determined the correlates we can compute the residuals too. When these developments are substituted in the condition equations (7.08) we obtain

$$\begin{aligned}
[\tfrac{aa}{p}] k_a + [\tfrac{ab}{p}] k_b + [\tfrac{ac}{p}] k_c + w_a &= 0 \\
[\tfrac{ab}{p}] k_a + [\tfrac{bb}{p}] k_b + [\tfrac{bc}{p}] k_c + w_b &= 0 \\
[\tfrac{ac}{p}] k_a + [\tfrac{bc}{p}] k_b + [\tfrac{cc}{p}] k_c + w_c &= 0
\end{aligned} \tag{7.12}$$

Also these equations are called *normal equations* although they have been constructed in a different way than equations (6.12). The number of these normal equations is r and the solution of them gives the correlates k_j, which, in their turn, give the residuals v by the aid of developments (7.11).

It is customary to compute the weighted square sum $[pvv]$ too. It gives a valuable check because it can be computed in several different ways, *e.g.* by formula

$$[pvv] = - k_a w_a - k_b w_b - k_c w_c \tag{7.13}$$

Furthermore, from it we obtain the standard error unit weight

$$m = \sqrt{\frac{[pvv]}{r}} \qquad (7.14)$$

7.2 *Instructions for practical computations*

1. *The sketch* is prepared in a similar way as for method A. In addition, the conditions are marked by a, b, c... or other symbols written close to the respective configurations as the triangles, center points of side conditions, the loops and the chains between fixed points. As to the order of conditions, those are taken first which imply only a few measured quantities or which have only +1 or -1 as coefficients, the long condition equations with non-integer coefficients are left to the last places.

2. *The fixed quantities* which are required for the conditions are compiled as in method A.

3. The list of *observations* is similar as in method A.

4. *The misclosures* are computed by the formulas (7.03) with such a precision that they contain two or three significant figures in most cases. It is important that all conditions which are independent of each other are found. As a check, a few conditions depending on the other may be added, *e.g.* the sum of all loops.

Some times the computation of misclosures requires several sheets with different schemes, some times they can be found immediately by the aid of a desk calculator. The coefficients a, b, c... of the condition equations are computed along with the misclosures on the same sheets; for them, however, an accuracy of three or four significant figures is sufficient.

5. *The condition equations* are summarized on a special table where for each equation one column, for each measured quantity one row is reserved. Only the coefficients are written in the scheme; this array forms then the matrix B as explained in Chapter 13.2. Most coefficients are = 0, these places are usually left empty. Often the significant coefficients are +1 or -1; these elements can be written in the right places by the aid of the sketch. All the other condition equations are multi-

plied or divided by such integers that their coefficients are also of order 1. An extra column is reserved for the row sums

$$s_i = a_i + b_i + c_i + \ldots$$

If the number of conditions is so great that several sheets are necessary, it is advisable to have a sum control column on each sheet.

If the observations have different weights, an additional column is reserved for the inverse weights $\frac{1}{p_i}$. If both the coefficients a, b, $c\ldots$ and the inverse weights are other figures than 0 or 1, each of such rows can be multiplied by $\frac{1}{p_i}$ and the products are written immediately on the next row.

6. *The normal equations* are formed by the same mechanical computation procedures as in method A; the only difference is that, instead of the product sums $[pal]$, $[pbl]$, $[pcl]\ldots$, the misclosures w_a, w_b, $w_c\ldots$ are written. After that, we need two columns for the sum controls. First, the product sums $[\frac{as}{p}]$, $[\frac{bs}{p}]$, $[\frac{cs}{p}]$ are computed and compared with the row sums without misclosures w, then, on each row, w is added to the row sum for the later sum controls. In the latter column S, on an extra row, the sum of misclosures can be written.

7. *The solution of the normal equations* will be explained in Section 8. The solution, numerical values of k_a, k_b, $k_c\ldots$ can be written on an additional row of scheme 5. The residuals v_i can now be computed by formulas (7.11) and written in an additional column of scheme 5. If the observations have different weights, it is advisable to write two columns, first one for the products $p_i v_i$ according to formula (7.10), second one for $v_i = \frac{1}{p_i} \cdot p_i v_i$.

For checking purposes, the product sum $[pvv]$ is computed and compared with the values obtained on sheet 7. Another check which never should be neglected is accomplished by writing the adjusted values $L_{a,i}$ into scheme 3 and computing the remaining misclosures by formulas (7.02).

In the method B, the computation of parameters is carried out first after the finished adjustment. The computations can be arranged as explained for step 5 of method A, starting from the adjusted values of the measured quantities and using the extreme precision. If the adjustment has been correctly performed, the same values of parameters must be obtained from all possible combinations of the observations.

8 *SOLUTION OF THE NORMAL EQUATIONS*

8.1 *The principle*

The solution of the normal equations consists of two parts, *elimination* and *back solution*. The former requires as many series of computations as there are unknown quantities to be solved (u in method A, r in method B) and it sets practical limits to the extension of nets which can be adjusted as one entity; these limits, however, are continuously widened by the development of electronic computers. The following instructions, nevertheless, are intended for the manual computations only at which either a slide rule or a desk calculator is used.

As a general principle, the elimination is carried out in such a way that the original set of N (u or r) equations with N unknowns is reduced to a new set of $N-1$ equations with $N-1$ unknowns, this again to a third set of $N-2$ equations with $N-2$ unknowns and so forth until only one equation with one unknown remains. In the back solution, the last unknown is solved from this single equation and substituted into one equation of each of the previous sets. Then the preceding unknown is solved and so backwards. The value of the unknown which was eliminated first will be obtained last.

In the general case of linear equations with unsymmetrical coefficients, an experienced person can find and advantageous order in which the unknowns should be eliminated. For the normal equations with symmetrical coefficients, the best order is usually found when the most frequently used parameters in method A or the longest condition equations in method B are taken on the last places. After that, other precautions hardly are profitable but some mechanical procedure with specially planned schemes should be used. All schemes which will be given in this Section are based on the *algorithm of Gauss* the most important feature of which is that all sets of equations after the elimination of each unknown in turn are similarly symmetrical as the original set. Then in each set the same procedure can be used: *the first equation of the set is subtracted from every other equation in turn but first it must be multiplied by such a factor that the first unknown disappears.* The first equation of each set plays, accordingly, a special role. Also the back solution is performed by the aid of them only. When a computing machine is available, it is not necessary to write down any other equations than these first equations of successive sets. We shall call them briefly *reduced equations.*

For the comprehensive learning of the practical applications of Gauss' algorithm, a student should perform the solution of one small set of normal equations using all the schemes explained below. For this purpose, we give detaled instructions with notations and formulas which are quite unnecessary after a few exercises.

8.2 *Original scheme for slide rule computations*

(Numerical example: Table III/9.1)

The elimination is performed exactly as explained above. Only the coefficients, the absolute terms, and the sum controls are written down but, because of the symmetry, the coefficients below the principal diagonal can be left out. All sets, therefore, are tapering downward; on the last row of each set there are two figures which do not represent any equations but belong to the checking computations. Above each set, an empty row must be reserved for the back solution.

For the instructions which will be given below we introduce the following notations. Each figure is represented by three symbols (h, i, j) representing an "address". The first symbol refers to the set; the original normal equations belong to set $h = 1$. The second symbol refers to the row; the first row of each set is $i = h$, the last row is always $i = N+1$, where N is the number of original equations. The empty rows to be filled during the back solution may be addressed by $i = 0$. The third symbol refers to the column. During the elimination there are $N+2$ columns; each row begins from column $j = i$. As headings of the columns, we can write $x, y, z \ldots \ell, S$ or $a, b, c \ldots w, S$ in addition or instead of the current numbers j.

By the aid of the addresses, we can write the general formula of elimination as follows

$$(h+1,i,j) = (h,i,j) - \frac{(h,h,i)}{(h,h,h)} (h,h,j) \qquad (8.01)$$

In other words: when we are computing row i for set $h+1$, the slide rule is put to show the ratio $\frac{(h,h,i)}{(h,h,h)}$; both these figures can be found from the first row of the previous set. By the aid of the third figure (h,h,j) on the same row, we can read from the slide rule the product represented by last term of equation (8.01). This product is immediately subtracted from (h,i,j) which can be found from the same column as the third factor. It is advisable to use four coins showing the four addresses which appear on the right hand side of (8.01). The first coin stays at (h,h,h) during the computation of the entire set $h+1$. The second coin stays at (h,h,i) during the computation of one row i and is moved then one step to right. Two other coins show (h,h,j) and (h,i,j) and move together step by step to right, starting for each row from the column of the second coin.

At the end of each row, a sum check is obtained in the same way as when forming the original normal equations. As a general feature, it must be noted that the coefficients on the principal diagonal are always diminishing from set to set but remain positive; if it is not so in practice, there must be an error somewhere.

In method A, there is an extra row with figures $[p\ell\ell]$ and $[p\ell s]$ in the set of original normal equations. After the elimination of the last unknown, the corresponding extra row shows two times the same figure which can be proved to be $[pvv]$; the proof will be given in Section 13. In the method B, we can write on the extra row after the normal equations 0 in the column of misclosures and $[w]$ in the second sum check column; after the elimination, this extra row shows two times the value of $-[pvv]$.

The back solution gives first the value of the last unknown:

$$x_N \quad \text{or} \quad k_N = -\frac{(N,N,N+1)}{(N,N,N)} \tag{8.02}$$

The general instruction of the back solution is, after this, that each time when a new unknown x_j has been found it is substituted in each of the first equations of the previous sets. For this purpose, we write products which can be denoted by addresses

$$(h,0,j) = (h,h,j)\ x_j \tag{8.03}$$

The preceding unknown x_{j-1} can now be computed by adding all the products on the row $(h=j-1,\ i=0)$ to the absolute term of the next row, $(h=j-1,\ i=1,\ j=N+1)$ and writing the sum above it:

$$(h,0,N+1) = (h,h,N+1) + \sum_{j=h+1}^{N} (h,0,j) \tag{8.04}$$

Then we have

$$x_h = -\frac{(h,0,N+1)}{(h,h,h)} \tag{8.05}$$

The back solution can be immediately checked by (6.13) or (7.13). The unknowns and the products of the check computations can be written in additional columns of the same scheme.

8.3 *Improved scheme for slide rule computations*

(Numerical example: Table VI/9.1)

If we review the entire development of the reduced equations which appear on the first row of each set of the previous scheme we realize that all subtractions according to formula (8.01) can be combined to one operation

$$(i,i,j) = (1,i,j) - \sum_{h=1}^{i-1} \frac{(h,h,i)}{(h,h,h)}\ (h,h,j) \tag{8.06}$$

Therefore, it is advisable to write down immediately the last terms of formula (8.01) without the immediate subtraction. In this way, we can avoid mistakes which easily might affect the sign and the position of the decimal points of these slide rule readings. These readings are collected in special columns for the summation according to formula (8.06).

As a further simplification, if one of the coefficients (h,h,i) is zero, we can just forget it but in the previous scheme we had to write one row unchanged into the new set again. Nevertheless, we can use the addresses defined for the previous scheme but with the alteration that for the first rows of each set, i is smaller than h. The general formula of computation reads:

$$(h,i,j) = -\frac{(i,i,h)}{(i,i,i)}\;(i,i,j) \qquad (8.07a)$$

For the last row of each set we have $i = h$ and

$$(h,h,j) = (1,h,j) + \sum_{i=1}^{h-1} (h,i,j) \qquad (8.07b)$$

The sets are no more tapering but each rectangular set is usually longer and narrower than the former sets. The only defect of the improved scheme is that the sum control is obtained easily for the last row of each set only.

The back solution is performed in the same order and by the same formulas as for the original scheme. The only difference is that the empty rows which can be filled during the back solution are now found below the reduced equations, not above.

8.4 *Scheme of Doolittle for desk calculators*

(Numerical example: Table VI/9.4)

When the product of two factors is formed by a desk calculator, it can be at the same time added to or subracted from an accumulating sum of all similar products. In formula (8.06), the products appearing in the last term can be formed in this way if the quotients $\frac{-(h,h,i)}{(h,h,h)}$ are computed and written down beforehand. In other words, each reduced equation must be immediately divided by it's first coefficient (or multiplied by the inverse of it); in addition, the sign is changed. We shall call these divided reduced equations briefly *end equations*.

In this scheme each end equation is written down on the next row after the corresponding reduced equation.

Anticipating our matrix notations, we denote the coefficients of the normal equations by n_{ij} (i referring to the row, j to the column), those of the reduced equations by r_{ij} and those of the end equations by t_{ij}. The first reduced equation is identical with the first normal equation. The general formulas for the other equations read

$$r_{ij} = n_{ij} + \sum_{h=1}^{i-1} t_{hi} r_{hj} \tag{8.08a}$$

$$t_{ij} = - \frac{r_{ij}}{r_{ii}} \tag{8.09a}$$

It is advisable to use three coins during the computation of each r_{ij}. The first coin remains at n_{ij} on the set of normal equations. The second coin moves along column i, the third along column j. One of the latter coins shows coefficient r, the other coefficient t on the next row. Because $t_{hi} r_{hj} = r_{hi} t_{hj}$ it is theoretically insignificant which coin is put on the upper, which on the lowe row but in order to reduce the errors due the rounding off, the two medium sized coefficients of these four should always be picked up, instead of the smallest and greatest ones.

After each row, the usual sum control can be performed. The pairs of rows get shorter and shorter until on an extra row the value of $[pvv]$ is obtained two times.

The back solution is carried out by the aid of the end equations only. The general formula of computation reads

$$x_i = t_{i,N+1} + \sum_{j=i+1}^{N} t_{ij} x_j \tag{8.10a}$$

The back solution can be checked by (6.13) or (7.13).

8.5 *Fillmatrix scheme*

(Numerical example: Table VII/9.6)

The method of computation is the same as for the scheme of Doolittle but the end equations are not written on rows but in columns. Because each $t_{ii} = -1$, it is not written at all but the rest of the coefficients t_{ij} are written in the column from r_{ii} downward. The scheme is not tapering downward but a square, the triangular parts of r and t can be separated from each other by a stairline, which goes next to left and below the coefficients r_{ii}. The computation formulas read now

$$r_{ij} = n_{ij} + \sum_{h=1}^{i-1} t_{ih} r_{hj} \qquad (8.08b)$$

$$t_{ji} = - \frac{r_{ij}}{r_{ii}} \qquad j > i \qquad (8.09b)$$

because, according to our general notation, the first subscript indicates the row, the second subscript the column. Each new coefficient r_{ij} requires a summation of products the first factor of which can be found from the same row i to left from the stairline, the second factor from the same column j of coefficients r_{ij} already computed.

The formula for the back solution reads

$$x_j = t_{N+1,j} + \sum_{i=j+1}^{N} t_{ij} x_i \qquad (8.10b)$$

Each unknown x_j requires a summation of products the first factor of which can be found from the same column j downward from the stairline, the second factor from the row of values x_j already computed.

8.6 *Scheme of Cholesky*

(Numerical example: Table VIII/16.7)

In the solution method of Cholesky, the reduced equations

are not written down immediately but each coefficient r_{ij} is first divided by the square root of the first coefficient on the same row:

$$k_{ij} = \frac{r_{ij}}{\sqrt{r_{ii}}} \tag{8.11}$$

Because in formula (8.08a) the product $t_{hi} r_{hj}$ now can be replaced by $k_{hi} k_{hj}$, it is not necessary to write double rows (or two triangles) for r and t but just one triangle for coefficients k. In this way some writing work can be saved but there will be essentially more of the computation work. First, the square root of each r_{ii} must be computed. Secondly, during the back solution, for each unknown a division by k_{ii} is required. Therefore, the author does not recommend this method for small manual computations. As to the programming of automatic computers, we may point out that only the coefficients r_{ij} are required in the inner storage because the terms with three factors in formula (8.06) can be computed and summed up very rapidly by these machines.

9. *ADJUSTMENT OF GEODETIC NETS*

9.1 *Levelling*

A net of spirit levelling is usually adjusted by method B (by correlates). It is possible, however, that a net of second order contains many fixed points of first order in which case the increased number of conditions renders method A less laborious.

During the net adjustment, we count as net points only those bench marks which already have fixed values of height and those new points from which at least three levelling lines lead to other net points. If there are no fixed points, one of the net points must be selected as the initial point with an arbitrary fixed height. Those new points from which only one or two lines have been levelled are not included into the net adjustment but connected or interpolated afterwards.

In method A, the heights of the new net points serve as parameters. The number of those points is denoted by u.

In method B, each closed loop gives a condition. In addition, each fixed point except one initial point gives another condition. If there are n levelled lines between the net points, the total number of conditions is $r = n - u$.

The weights of the levelled lines are usually supposed to be inversely proportional to the lengths s of the lines; in method B we set the inverse weight $\frac{1}{p} = s$ in kilometers. If some lines with exceptional weights (*e.g.* those going through a rough terrain) have been included in the net to be adjusted, these weights must be duly estimated; often the weak lines have been levelled several times and the mean taken with the normal weight.

The net of precision levelling of Finland is tied to mareographs with continuous records of the sea level; the mean of recorded heights of the sea level serves as the initial point. Special attention must be directed to the land upheaval which is supposed to be different at each point but constant during the period of levellings. Therefore, each net point needs two parametres, one showing the height at a given epoch, the other showing the annual rise. An exact adjustment without interpolations is possible only for the lines which have been levelled two times with a long time interval.

When a local net of levelling is connected to this primary net, the heights of the fixed points should be reduced to another epoch which should be set in the future because the complete survey and building of the area will take a long time. The height differences of the new levelling are usually not reduced according to the established tilting of the ground because the local adjustment itself signifies an interpolation between the fixed points.

As a *numerical example* of the adjustment of local nets, we take a part of the levelling of Helsinki. The net of the down town area has been measured with an extraordinary precision. It will be adjusted here by method B using only one fixed point of the fundamental net of Finland. The net of the northern suburbs is adjusted by method A between the city net and another fixed point.

Both nets are represented by the sketch in fig. 1 and by the list of observed height differences in Table I. The points A-12 belong to the city net, the points 13-B to the suburb net. The lines have been identified by current numbers ν and by arrows showing the direction of height differences (end point minus starting point). The loops of the city net are marked by symbols a - g referring to the conditions; each condition is formed clockwise around the loop. The length of each line is given by s in kilometers, the weights $p = \frac{1}{s}$ are given in the scheme of observation equations for the suburb net only.

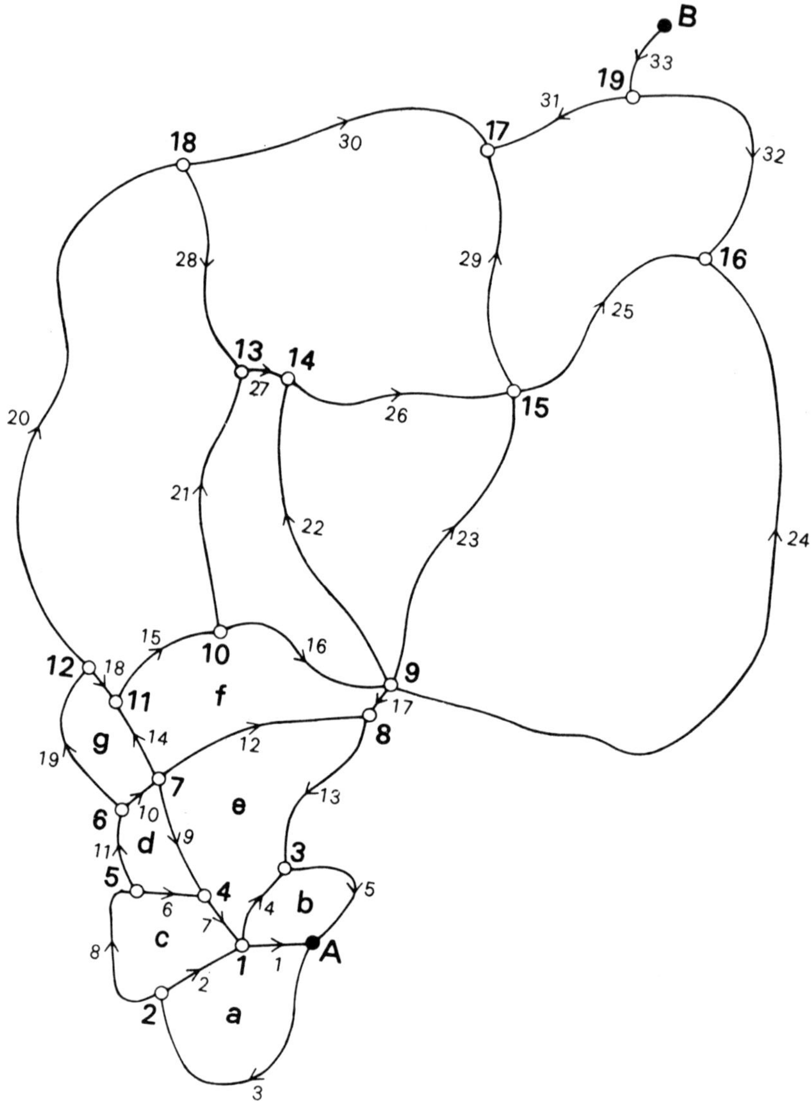

Fig. 1. *Part of the levelling net of Helsinki*

Table I *List of observations*

V	*P* → *P*		*Observed*	*s*	*Adjusted*	
			City net			
			m		*m*	
1	1	A	- 5 496.56	0.80	- 5 496.51	
2	2	1	4 873.75	0.83	4 874.04	
3	A	2	0 623.23	3.45	0 622.47	
4	1	3	- 2 399.67	1.05	- 2 399.88	
5	3	A	- 3 096.26	1.30	- 3 096.63	
6	5	4	- 5 092.72	0.58	- 5 093.08	
7	4	1	0 893.94	0.70	0 893.60	
8	2	5	9 074.47	1.70	9 073.50	
9	7	4	1 575.79	1.30	1 575.97	
10	6	7	- 8 976.21	0.48	- 8 976.38	
11	5	6	2 307.29	0.83	2 307.33	
12	7	8	11 379.65	2.29	11 379.25	
13	8	3	-11 309.37	2.20	-11 309.56	
14	7	11	- 0 223.22	0.74	- 0 223.45	
15	11	10	2 992.52	1.20	2 992.62	
16	10	9	9 803.66	2.11	9 803.84	
17	9	8	- 1 193.76	0.15	- 1 193.75	
18	12	11	- 5 288.10	0.30	- 5 287.98	
19	6	12	- 3 912.66	2.00	- 3 911.86	
			Suburb net		*Approximate*	*ℓ*
20	12	18	14 179.00	7.40	14 179.00	0
21	10	13	12 453.64	3.22	12 453.64	0
22	9	14	9 799.60	3.92	9 799.60	0
23	9	15	- 7 388.35	3.50	- 7 388.35	0
24	9	16	-14 767.80	10.47	-14 783.83	-16.03
25	15	16	- 7 391.58	2.16	- 7 395.48	- 3.90
26	14	15	-17 188.58	2.45	-17 187.95	0.63
27	13	14	7 156.45	0.63	7 149.80	- 6.65
28	18	13	- 4 010.33	2.14	- 4 020.72	-10.39
29	15	17	14 252.58	2.43	14 257.30	4.72
30	18	17	0 216.03	2.76	0 198.43	-17.60
31	19	17	+15 848.19	1.82	15 848.19	0
32	19	16	- 5 804.59	3.45	- 5 804.59	0
33	B	19	- 7 334.19	1.11	- 7 334.19	0

Table II *Condition equations of the city net*

v	*a*	*b*	*c*	*d*	*e*	*f*	*g*	*s*	*pv*	*v*	*pvv*
1	1	-1						0.80	+0.065	+0.05	0.003
2	1		-1					0.83	+0.349	+0.29	0.101
3	1							3.45	-0.220	-0.76	0.167
4		1			-1			1.05	-0.197	-0.21	0.041
5		1						1.30	-0.285	-0.37	0.105
6			1	-1				0.58	-0.621	-0.36	0.224
7			1		-1			0.70	-0.481	-0.34	0.161
8			1					1.70	-0.569	-0.97	0.550
9				1	-1			1.30	+0.140	+0.18	0.025
10				1			-1	0.48	-0.349	-0.17	0.058
11				1				0.83	+0.052	+0.04	0.002
12					1	-1		2.29	-0.173	-0.40	0.068
13					1			2.20	-0.088	-0.19	0.017
14						1	-1	0.74	-0.316	-0.23	0.074
15						1		1.20	+0.085	+0.10	0.009
16						1		2.11	+0.085	+0.18	0.015
17						1		0.15	+0.085	+0.01	0.001
18							1	0.30	+0.401	+0.12	0.048
19							1	2.00	+0.401	+0.80	0.322
										[pvv]	1.991

The coefficients of both the condition equations and the observation equations are +1 or -1 (or zero) which can easily written in the respective schemes by the aid of the sketch, see Tables II and V. Using the coefficients of Table II and the obsered values of Table I, the misclosures can be computed immediately and written in Table III of normal equations, column w, rows 1,1 - 1,7. For the adjustment of the suburb net, the approximate heights of the points 13-19 have been computed by the aid of the results of the adjustment of the city net and using those observed differences which follow the shortest way from the fixed points, see Table IV. The other observed height differences have been subtracted from the approximate differences and the discrepancies have been written in column ℓ of Table V.

The computation of the coefficients of the normal equations consists of picking up values of p or s, see Tables III and VI. The solution of the normal equations has been controlled by column S. The original scheme for slide rule computations has been used

Table III *Normal equations and the solution*

h,i j	a 1	b 2	c 3	d 4	e 5	f 6	g 7	w 8	S 9	k 0	-kw
1.0		+0.23	+0.47					+0.42	+1.12	-0.220	0.09
1.1	5.08	-0.80	-0.83					+0.42	+3.87		
1.2		3.15			-1.05			+0.63	+1.93		
1.3			3.81	-0.58	-0.70			+1.94	+3.64		
1.4				3.19	-1.30		-0.48	-0.41	+0.42		
1.5					7.54	-2.29		+0.22	+2.42		
1.6						6.49	-0.74	-0.45	+3.01		
1.7							3.52	-1.33	+0.97		
1.8								0	+1.02		
2.0			+0.07		+0.09			+0.70	+0.86	-0.285	0.20
2.2		3.02	-0.13		-1.05			+0.70	+2.54	0	
2.3			3.67	-058	-0.70			+2.01	+4.27	0	
2.4				3.19	-1.30		-0.48	-0.41	+0.42		
2.5					7.54	-2.29		+0.22	+2.42		
2.6						6.49	-0.74	-0.45	+3.01		
2.7							3.52	-1.33	+0.97		
2.8								-0.03	+0.70	+1	
3.0				-0.03	+0.07			+2.04	+2.08	-0.569	1.10
3.3			3.66	-0.58	-0.75			+2.04	+4.38	-1	
3.4				3.19	-1.30		-0.48	-0.41	+0.42		
3.5					7.17	-2.29		+0.46	+3.30	-1.	
3.6						6.49	-0.74	-0.45	+3.01		
3.7							3.52	-1.33	+0.97		
3.8								-0.19	+0.12	0	
4.0					+0.12		-0.19	-0.09	-0.16	+0.052	0.02
4.4				3.10	-1.42		-0.48	-0.09	+1.11	0	
4.5					7.02	-2.29		+0.88	+4.19	0	
4.6						6.49	-0.74	-0.45	+3.01		
4.7							3.52	-1.33	+0.97		
4.8								-1.33	-2.32	0	
5.0						-0.19	-0.09	+0.84	+0.56	-0.088	0.02
5.5					6.37	-2.29	-0.22	+0.84	+4.70	0	
5.6						6.49	-0.74	-0.45	+3.01		
5.7							3.45	-1.34	+1.14	+1	
5.8								-1.33	-2.29	+1	
6.0							-0.33	-0.15	-0.48	+0.085	0.04
6.6						5.67	-0.82	-0.15	+4.70	0	
6.7							3.44	-1.31	+1.31	0	
6.8								-1.44	-2.90	0	
7.0								-1.33	-1.33	+0.401	0.53
7.7							3.32	-1.33	+1.99	0	
7.8								-1.44	-2.78	+1	
8.8								-1.97	-1.97	0	2.00

Table IV *List of the heights of the net points*

City net		Suburb net			
P	Adjusted	P	Approximate	Δh	Adjusted
	m		m		m
A	4 453.60	13	22 703.35	- 2.27	22 701.08
1	9 950.11	14	29 853.15	+ 3.78	29 856.93
2	5 076.07	15	12 665.20	+ 3.17	12 668.37
3	7 550.23	16	5 269.72	+ 6.20	5 275.92
4	9 056.51	17	26 922.50	+ 1.65	26 924.15
5	14 159.58	18	26 724.07	-12.13	26 711.94
6	16 456.92	19	11 074.31	+ 1.55	11 075.86
7	7 480.54	B	18 408.50		
8	18 859.79				
9	20 053.55				
10	10 249.71				
11	7 257.09				
12	12 545.07				

Table V *Observation equations of the suburb net*

V	13	14	15	16	17	18	19	ℓ	p	pℓ	v	pvv
20						1		0	0.135		-12.13	19.86
21	1							0	0.311		- 2.27	1.60
22		1						0	0.255		+ 3.78	3.64
23			1					0	0.286		+ 3.17	2.87
24				1				-16.03	0.096	- 1.539	- 9.83	9.28
25			-1	1				- 3.90	0.463	- 1.806	- 0.87	0.35
26		-1	1					+ 0.63	0.408	+ 0.257	+ 0.02	0
27	-1	1						- 6.65	1.587	-10.554	- 0.60	0.57
28	1					-1		-10.39	0.467	- 4.852	- 0.53	0.13
29			-1		1			+ 4.72	0.412	+ 1.945	+ 3.20	4.22
30					1	-1		-17.60	0.362	- 6.371	- 3.82	5.28
31					1		-1	0	0.549		+ 0.10	0
32				1			-1	0	0.290		+ 4.65	6.27
33							1	0	0.901		+ 1.55	2.16

$[pvv] = 56.23$

for the city net (Table III), the improved scheme for the suburb net (Table VI).

In Table III, the correlates k have been checked by computing $[pvv] = -[kw]$. After that, the columns pv and v in Table II have been filled up. As a check, $[pvv]$ has been computed once more. Finally the definitive heights of points 1 - 12 have been computed and written in Table IV.

Table VI. *Normal equations and the solution*

h, i / j	13 / 1	14 / 2	15 / 3	16 / 4	17 / 5	18 / 6	19 / 7	l / 8	S / 9		Δh / 0
1,1	2,365	−1,587				−0,467		+ 5,701	+ 6,012		
1,0		−6,004				+5,663		+ 5,701	+ 5,360		− 2,27
1,2		2,250	−0,408					−10,811	−10,556		
2,1		−1,065				−0,313		+ 3,826	+ 4,034		
2,2		1,185	−0,408			−0,313		− 6,985	− 6,522	+1	
2,0			−1,293			+3,795		− 6,985	− 4,483		+ 3,78
1,3			1,569	−0,463	−0,412			+ 0,118	+ 0,404		
3,2			−0,140			−0,108		− 2,405	− 2,245		
3,3			1,429	−0,463	−0,412	−0,108		− 2,287	− 1,841	0	
3,0				−2,869	−0,681	+1,310		− 2,287	− 4,527		+ 3,17
1,4				0,849			−0,290	− 3,345	− 3,249		
4,3				−0,150	−0,133	−0,035		− 0,741	− 0,596		
4,4				0,699	−0,133	−0,035	−0,290	− 4,086	− 3,845	0	
4,0					−0,220	+0,424	−0,449	− 4,086	− 4,331		+ 6,20
1,5					1,323	−0,362	−0,549	− 4,427	− 4,427		
5,3					− 119	− 31		− 659	− 531		
5,4					− 25	− 7	− 55	− 777	− ·732		
5,5					1,179	−0,400	−0,604	− 5 863	− 5,690	+2	
5,0						+4,850	−0,936	− 5,863	− 1,949		+ 1,65
1,6						0,964		+11,223	+11,358		
6,1						− 92		+ 1,126	+ 1,187		
6,2						− 83		− 1,845	− 1 722		
6,3						− 8		− 173	− 139		
6,4						− 2	− 15	− 204	− 193		
6,5						− 136	− 205	− 1,989	− 1,930		
6,6						0,643	−0,220	+ 8,138	+ 8,561	0	
6 0							−0,341	+ 8,138	+ 7,797		− 12,13
1,7							1,740	0	+ 0,901		
7,4							− 120	− 1,695	− 1.595		
7,5							− 309	− 3,004	− 2,914		
7,6							− 75	+ 2,784	+ 2,929		
7,7							1,236	− 1,915	− 0,679	0	+ 1,55
1,8								273,78	272,24		273,78
8,1								− 13,74	− 14,49		− 12,92
8,2								− 41,17	− 38,44		− 40,90
8,3								− 3,66	− 2,95		+ 0,37
8,4								− 23,88	− 22,48		− 20,73
8,5								− 29,16	− 28,29		− 7,32
8,6								−103,00	−108,35		−136,09
8,7								− 2,97	− 1,05		0
8,8								56,20	56,19		56,19

In the suburb net, the solution of the normal equations gives corrections Δh to the approximate heights of the new points. The results are rewritten in Table IV. As a check, $[pvv]$ has been computed three times in Table VI and once more in Table V.

The standard error of unit weight in the city net is

$$m = \sqrt{\frac{1.99}{7}} = \sqrt{0.284} = \pm 0.53 \text{ mm}/\sqrt{\text{km}}$$

and in the suburb net

$$m = \sqrt{\frac{56.2}{7}} = \sqrt{8.03} = \pm 2.83 \text{ mm}/\sqrt{\text{km}}$$

9.2 *Adjustment of triangulations, summary*

The mathematical treatment of a triangulation involves several adjustments. In this book, these have been scattered in various chapters. Thérefore we shall give here a systematic summary.

Preliminary adjustments are smaller operations by the aid of which a group of measurements is replaced by a few mathematical functions, "fictitious observations"; in the later adjustments these functions serve as observed quantities. Therefore it. is necessary to compute the variances, some times also the covariances of these functions. As typical preliminary adjustments, we may mention *station adjustments* and the *expansion nets of base lines*. From the former, we have examples in Chapters 4, 3, (Station adjustment with complete sets), 16.1 (Station adjustment with incomplete sets) and 5.5 (Measurement of angles in all combinations), from the latter, in Chapter 16.2.

The *definitive net adjustments* can be divided into many classes. The *space triangulations* by the aid of artificial satellites or air born flashes require an adjustment in three dimensions. These triangulations cover very large areas; a world wide triangulation around the entire earth is still a problem of the future. The classical *triangulations of the first order* are adjusted along the surface of a reference ellipsoid which is usually determined beforehand. The deflections of the vertical, or the astronomical determinations of latitude, longitude and especially those of the azimuth, can be included into the same adjustment; then also the best fitting ellipsoid could be found. The gravimetric observations and the measurements of vertical angles have not been included in these classical adjustments. Because all these adjustments deal with an immense number of observations of various kinds, we can not go into the details in this book.

Also the *city triangulations* are divided into classes of different orders. In Chapter 9.3, we shall give an example of the adjustment of the *principal net*. Such a net requires usually it's own base line; the accuracy of the sides of the state triangulation seldom is sufficient for a city survey. The principal nets are usually adjusted by correlates (method B).

The *second order nets* are adjusted inside the loops or triangles of the first order or principal nets the points of which are considered fixed. In Chapter 16.3 we shall have an example by method B. Modern triangulations and trilaterations in which many sides have been measured by geodimeters or tellurometers are adjusted preferably by method A; in Chapter 9.5 we shall have an example. After the definitive adjustment of the second order nets, those of the third order etc. are dealt with. Eventually, *single points* are connected to these nets; an example will be given in Chapter 9.7.

After the definite adjustments of a triangulation, there still can be cases of a *readjustment*. *E.g.*, if an ancient triangulation is revised in such a way that some of the points have been measured again with a higher precision, the rest of the old coordinates can be reoriented into the new system by a *coordinate transformation* which is described in Chapter 16.6. The same problem arises when two isolated triangulations expand and meet each other.

9.3 *Adjustment by correlates of a city triangulation*

In a normal case, only the bearings (or the angles) are used as measured quantities. Therefore, two points must be considered as fixed. One point may have the coordinates of the state triangulation or, in the lack of these, just arbitrary round values in such a way that no point of the net receives negative coordinates. The second point is fixed as soon as there is a measured base line and an astronomical azimuth. The latter can be derived from the state triangulation, the former should be acquired by special measurements of highest precision. The coordinates x and y of all other points serve as parameters of the adjustment. For each mean series of bearings, the zero bearing serves as an additional parameter.

Suppose that at each point the bearings have been observed in one palmate set only. Then the number of zero bearings is P = the number of points and the number of unknown coordinates is $2(P-2)$. All told, there are $u = 3P - 4$ parameters.

Suppose that there are S sides in the net and that the bearing has been observed at both ends of each side. Then the number of observations is $n = 2S$ and that of the conditions

$$r = n - u = 2S - 3P + 4 \qquad (9.01)$$

These conditions are of two different types. *Triangle equations* express the condition that the sum of angles in each triangle — when reduced on the projection plane — should be two right angles. These equations can be replaced by other angle sum equations which can be obtained from the triangle equations by addition or by subtraction but the typical feature of these equations is that the coefficients are +1 or -1 (or 0). However, all possible equations which could be formed with this simple type are not independent of each other. The second type of conditions consists of *side equations* the coefficients of which are not integers. Also these equations can be formed in many ways but because they are more laborious for the numerical computations we shall take the minimum number of them. Now the smallest number of sides which are necessary between P points in order to form a consistent net of triangles is $2P - 3$ (one side between the two first points and two sides for each additional point).

Each side over this amount gives a side condition. Therefore, the number of independent side conditions is

$$r_2 = S - 2P + 3 \qquad (9.02)$$

The remaining number

$$r_1 = S - P + 1 \qquad (9.03)$$

of conditions can be written as independent angle sum equations. For each net configuration to be adjusted, it is necessary to check these numbers carefully because, otherwise, some conditions may be overlooked and the adjustment would fail.

The condition equations can be formed by the aid of a special scheme in form of a *list of triangles*. In this list, we reserve for each triangle four rows, three for the angles and one for the misclosure. The first column is reserved for the symbol of the triangle, (a, b, c, ...), the second column for the symbols of the corner points P, the third column for the symbols of the measured angles or, preferably, for the symbols of the bearings the difference of which forms the angle. By the aid of these symbols, the measured bearings can be found from the list of observations and the difference written in the fourth column. The misclosure can now be computed and written on the extra row.

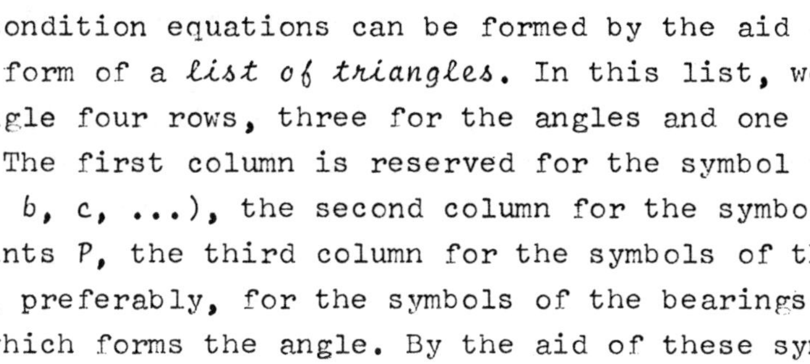

Fig.2. *Single chain of triangles*

The general structure of the side equations can be understood when we take first the case of a *single chain* (fig.2). Let s_0 be the length of the initial side , s_k the length of the end side.

In each triangle, we can name a forward angle α, a sideward angle β and a backward angle γ in such a way that α faces the next connecting side or the end side, β the border side of the chain and γ the initial side or the preceding connecting side. When s_o is multiplied by each sinα and divided by each sin γ through the chain, a value for s_k is found. If both s_o and s_k are fixed, the observed values of the angles reveal a misclosure w. Suppose that the computations is performed by the aid of logarithms. Then we can write

$$\begin{aligned} w &= \log\sin\alpha_1 + \log\sin\alpha_2 + \ldots + \log\sin\alpha_k \\ &- \log\sin\gamma_1 - \log\sin\gamma_2 - \ldots - \log\sin\gamma_k \\ &+ \log s_o - \log s_k \end{aligned} \tag{9.04}$$

As the unit of the misclosure, we take the sixth decimal place of the logarithm. Each time when "log sin" is interpolated from the logarithmic tables, we write down also the difference d_α or d_γ which shows the variation of the sixth decimal of the logarithm for the increase of 1" of the angle α or γ.

Denoting the corrections of angles by $v''_\alpha - v'_\alpha$ and $v''_\gamma - v'_\gamma$, the linearized condition equations can be written in form

$$\sum_1^k \left\{ d_\alpha (v''_\alpha - v'_\alpha) - d_\gamma (v''_\gamma - v'_\gamma) \right\} + w = 0 \tag{9.05}$$

where the summation includes all triangles between s_o and s_k.

Instead of logarithms, the computations can be performed by machines too. In this case, the formula of the misclosure reads

$$w = \rho \left(\frac{\sin\alpha_1 \sin\alpha_2 \ldots \sin\alpha_k s_o}{\sin\gamma_1 \sin\gamma_2 \ldots \sin\gamma_k s_k} - 1 \right) \tag{9.06}$$

where ρ = 206265" or = 636620"; the unit of the misclosure is then second of arc, sexagesimal or centesimal, respectively. Each time when sin α or sin γ is interpolated from numerical trigonometric tables with 7 or 8 decimal places, also the value of cot α or cot γ is written down with 3 or 4 decimal places. When d_α and d_γ are replaced by these cotangents, formula (9.05) is valid for this case too.

If the single chain forms a closed loop in such a way that

side s_k is the same as s_o, these sides disappear from formulas (9.04) and (9.06).

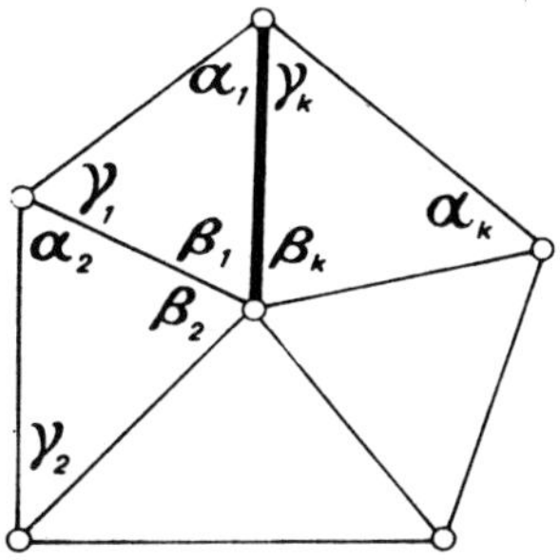

Fig. 3. *Central system*

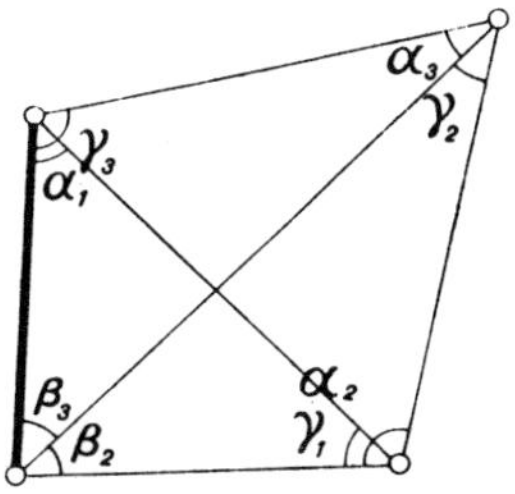

Central Point
$\beta_1 = \beta_2 + \beta_3$

Fig. 4. *Quadrilateral with two diagonals*

A *central system* (fig. 3) is a special but very common case of a loop where the inside border of the loop consists of a point. All sideward angles β are around the central point. If the palmate series of the connecting sides is circumscribed counterclockwise, the forward angles α form the right hand part, the backward angles γ the left hand part of each angle at the periphery of the figure.

A *quadrilateral* with two diagonals (fig. 4) can be regarded as a central system where the intersection point of the diagonals serves as the central point. There are no measured angles at this point but these sideward angles do not play any role in the side conditions. However, the formulas for the misclosure and the side equation are shorter if any one of the four observation points is regarded as the central point. It is recommended to use that point which leaves the largest triangle between the three remaining points.

For the formation of the side conditions, a special scheme

can be used; cfr. Table III/9.4. As an alternative, we can use additional columns in the list of triangles; cfr. Table IIa/16.2. In the latter scheme, the numerical values of sin or log sin are written in the fifth column, those of d_α and d_γ in the sixth column. Then for each side equation a narrow column is reserved. In this column, the forward angles are shown by sign +, the backward angles by sign -.

The scheme of the condition equations gives a summary of the coefficients. When writing down the coefficients of the side equations, the following instructions can be used. Each of the values of d must be brought in two places, on the row of v'' as well as of v'. The sign is changed for bearings v' and for backward angles γ. Because the same bearing can appear in two angles, the corresponding values of d must be added in the summary scheme. In order to make the absolute values of coefficients, in average, close to 1, each side equation can be divided by a small integer; *e.g.* by 2 when sexagesimal seconds and logarithms are used.

In an exceptional case, the angle appearing in a side equation is not measured. Such an angle must be expressed by the aid of the two other angles of the same triangle and the coefficient d must be written in four places of the scheme. If this exceptional case occurs in a quadrilateral, the point at which no angles have been measured can be used as the center without any complications.

Exercises:

1. A triangulation net contains five stations of which two are supposed to be fixed. At each station, all other stations have been observed in one theodolite set. How many side conditions and how many angle sum conditions must be taken into consideration for the net adjustment?

Answer: Three side conditions and six angle sum conditions.

2. In a quadrilateral ABCD, all six sides S_{AB}, S_{AD} etc. but no angles have been measured. Supposing that one point and one direction are fixed, form the condition equation for the adjustment of the net.

Solution: Compute the four great angles at the corner points, using formula (3.42). The misclosure w is = the sum of angles - 360°. The formulas for the coefficients are

$$\frac{\delta w}{\delta s_{AB}} = - \frac{s_{AB} - s_{AD} \frac{\cos(DAB)}{\sin(DAB)}}{s_{AB} s_{AD}} - \frac{s_{AB} - s_{BC} \frac{\cos(ABC)}{\sin(ABC)}}{s_{AB} s_{BC}} = - \frac{1}{k_{AB}} \quad \text{etc.}$$

$$\frac{\delta w}{\delta s_{AC}} = \frac{s_{AC}}{s_{AB} s_{BC} \sin(ABC)} + \frac{s_{AC}}{s_{CD} s_{AD} \sin(CDA)} = \frac{1}{k_B} + \frac{1}{k_D} \quad \text{etc.}$$

where k_B means the normal from point B to the side AC, k_{AB} the normal from the intersection point of AC and BD to side AB a.s.o.

Note: Also the coördinates of the points could be used but the areas of the triangles are not recommended because both the misclosure and the coefficients could then be close to zero. Regardless of the method of computation, the misclosure and the coefficients must have the same proportions to each other.

9.4 *Numerical example*

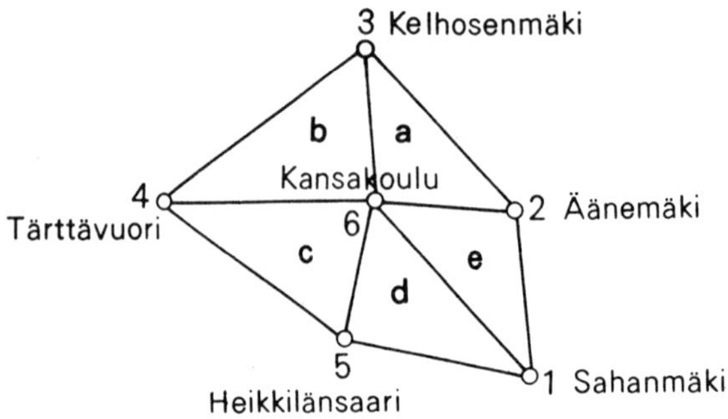

Fig. 5. *Principal net of a city triangulation*

Fig. 5 represents a typical case of the principal triangulation of small city. There are $P = 6$ points and $S = 10$ sides with observed bearings at both ends. According to formulas (9.02) and (9.03), we must form one side equation and 5 angle sum equations. The latter are denoted a, b, c, d and e, the side equation (central system) by f.. The bearings are denoted by two numbers, *e.g.* 15 means "from point 1 to point 5". These symbols are in columns V of Tables I - IV.

Table I gives the observed values only. According to the usual practice, the first bearing at each point is taken as zero; the true value of this zero bearing is an unknown parameter which does not play any role in the adjustment by correlates. All bearings are supposed to have equal weights $p = 1$.

Table II gives, in addition to the relevant symbols, the measured values of the angles. By the aid of this table, we can read the angle sum equations, *e.g.* the condition a reads

$$v_{23} - v_{26} + v_{62} - v_{63} + v_{36} - v_{32} - 7''08 = 0$$

Of course, only the coefficients +1 or -1 are required; they are brought directly into Table IV.

The misclosure of the side equation has been computed by formula

$$\omega_f = \log \frac{\sin(23-26)\sin(34-36)\sin(45-46)\sin(51-56)\sin(12-16)}{\sin(36-32)\sin(46-43)\sin(56-54)\sin(16-15)\sin(26-21)}$$

using Table III in which the differences *d* are visible too. These differences, as well as the misclosure, have been divided by 2 before entering them into Table IV.

In Tables IV-VI, there are additional columns for the sum control *S* and for the products $\sigma\sigma$ and σS. These latter are necessary, of course, only when a slide rule is used; if a desk calculator is available, the products are formed by accumulative multiplications. The columns v and vv have been computed after the adjustment.

Table I
List of bearings

V	Observed
15	0°00'00"00
16	37 49 38.10
12	75 03 30.05
21	0 00 00.00
26	98 59 21.61
23	140 35 53.20
32	0 00 00.00
36	40 26 47.53
34	98 04 53.10
43	0 00 00.00
46	35 46 11.78
45	73 06 36.15
54	0 00 00.00
56	63 38 27.65
51	153 09 10.93
63	0 00 00.00
62	97 56 33.80
61	141 43 17.73
65	194 22 56.82
64	273 24 11,66

Table II
List of triangles

Δ	P	V	Obs.angle
a	Ää	23 -26	41°36'31"59
	Ka	62 -63	97 56 33.80
	Ke	36 -32	40 26 47.53
			ω - 7.08
b	Ke	34 -36	57 38 05.57
	Ka	63 -64	86 35 48.34
	Tä	46 -43	35 46 11.78
			ω + 5.69
c	Tä	45 -46	37 20 24.37
	Ka	64 -65	79 01 14.84
	He	56 -54	63 38 27.65
			ω + 6.86
d	He	51 -56	89 30 43.28
	Ka	65 -61	52 39 39.09
	Sa	16 -15	37 49 38.10
			ω + 0.47
e	Sa	12 -16	37 13 51.95
	Ka	61 -62	43 46 43.93
	Ää	26 -21	98 59 21.61
			ω - 2.51

Table III *Side equation*

Δ	*Forward angles* V	log sin	diff	*Backward angles* V	log sin	diff
a	23-26	.822194.72	2.37	36-32	.812069.46	2.47
b	34-36	.926678.84	1.34	46-43	.766808.33	2.92
c	45-46	.782863.04	2.76	56-54	.952322.44	1.04
d	51-56	.999984.24	0.02	16-15	.787660.77	2.71
e	12-16	.781777.87	2.77	26-21	.994632.71	-0.33
σ	sum	+.313498.71			-.313493.71	
	ω	+5.00				

Table V does not show the first column for sum control (without w). The column $-kw$ has been computed after the solution of the normal equations, which has been carried out using the Doolittle scheme.

As a check, [vv] has been computed four times in Tables IV-VI. The standard error of one measured bearing is

$$m = \sqrt{\frac{27.1}{6}} = \sqrt{4.51} = \pm\ 2''12$$

Table IV *Condition equations*

V	a	b	c	d	e	f	S	ff	fS	v	vv
15				-1		+1.36	+0.36	1.85	0.49	+0.37	0.14
16				+1	-1	-2.74	-2.74	7.51	7.51	-0.91	0.83
12					+1	+1.38	+2.38	1.90	3.28	+0.54	0.29
21					-1	-0.17	-1.17	0.03	0.20	-0.62	0.38
26	-1				+1	-1.02	-1.02	1.04	1.04	-0.29	0.08
23	+1					+1.19	+2.19	1.42	2.61	+0.90	0.81
32	-1					+1.23	+0.23	1.51	0.28	-1.06	1.12
36	+1	-1				-1.90	-1.90	3.61	3.61	+2.30	5.29
34		+1				+0.67	+1.67	0.45	1.12	-1.24	1.54
43		-1				+1.46	+0.46	2.13	0.67	+1.10	1.21
46		+1	-1			-2.84	-2.84	8.07	8.07	+0.65	0.42
45			+1			+1.38	+2.38	1.90	3.28	-1.75	3.06
54			-1			+0.52	-0.48	0.27	-0.25	+1.63	2.66
56			+1	-1		-0.53	-0.53	0.28	0.28	-1.17	1.36
51				+1		+0.01	+1.01	0.00	0,01	-0.46	0.21
63	-1	+1								-2.18	4.74
62	+1				-1					+0.35	0.12
61				-1	+1					+1.08	1.17
65			-1	+1						+1.20	1.44
64		-1	+1							-0.47	0.22
Σ								31.97	32.20		27.09

Table V *Normal equations*

j	1	2	3	4	5	6	w	s	-kw
i= 1	6.00	-2.00			-2.00	-0.92	-7.08	-6.00	6.95
2		6.00	-2.00			-1.73	+5.69	+5.96	6.79
3			6.00	-2.00		+3.17	+6.86	12.03	11.39
4				6.00	-2.00	-3.56	+0.47	-1.09	0.22
5					6.00	+3.27	-2.51	+2.76	1.57
6						31.97	+2.50	34.70	0.16
s							0,00	+5.93	27.08

Table VI *Doolittle scheme of solution*

j	1	2	3	4	5	6	w	s	k
i = 1	6.00	-2.00			-2.00	-0.92	-7.08	-6.00	
	-1.000	+0.333			+0.333	+0.15	+1.180	+1.000	+0.981
2		5.33	-2.00		-0.67	-2.04	+3.33	+3.95	
		-1.000	+0.375		+0.125	+0.382	-0.625	-0.743	-1.194
3			5.25	-2.00	-0.25	+2.41	+8.11	13.52	
			-1.000	+0.381	+0.048	-0.459	-1.545	-2.575	-1.661
4				5.24	-2.10	-2.64	+3.56	+4.06	
				-1.000	+0.401	+0.504	-0.679	-0.774	-0.460
5					4.40	+1.76	-2.64	+3.52	
					-1.000	-0.400	+0.600	-0.800	+0.626
6						27.90	+1.79	29.70	
						-1.000	-0.064	-1.064	-0.064
						[vv] =	27.08	27.08	

9.5 *Adjustment by coordinates*

Method A, adjustment by coordinates, instead of method B, adjustment by correlates, is to be preferred for the computation of triangulations in the following cases:

- in the nets of lower order when there are many fixed points of higher order,
- when there are many side equations (diagonals and central systems),
- when there are many measured sides (geodimeter or tellurometer observations).

The deciding factor is, as a general rule, the number of normal equations: when r = number of conditions turns out to be greater than u = number of parameters, method A is preferable. For the comparison of these numbers, we must compute $r = n - u$ where n = number of observed bearings and sides, u comprises two coordinates for each new point and one zero bearing for each series (*i.e.* mean of observed series).

For the adjustment by coordinates, the approximate values of the coordinates of the new points must be computed first. We shall not go into the details of this part of the work. The next part consists of computation of the observed bearings t and sides s on the basis of the fixed coordinates of the old points and of the approximate coordinates of the new points. On the plane of the accepted map projection, each side is computed in one direction only:

$$\tan t_{AB} = \frac{y_B - y_A}{x_B - x_A} \tag{9.07}$$

$$s_{AB} = \sqrt{(x_B - x_A)^2 + (y_B - y_A)^2} \tag{9.08}$$

These computations should be carried out with 7 or 8 significant figures. At the same time, for each side the following coefficients must be computed with 3 or 4 significant figures:

$$a_{AB} = -(y_B - y_A)\frac{\rho}{s_{AB}^2} \qquad (9.09)$$
$$b_{AB} = +(x_B - x_A)\frac{\rho}{s_{AB}^2}$$

for each observed bearing t_{AB} or t_{BA},

$$c_{AB} = \frac{x_B - x_A}{s_{AB}} \qquad (9.10)$$
$$d_{AB} = \frac{y_B - y_A}{s_{AB}}$$

for each measured side s_{AB}. For the derivation of these formulas, cfr.(3.36). These coefficients should be of order 1, therefore, when the unit of angles is 1", the unit of sides and coordinate differences should be 1 cm in city triangulations and 1 dm in state triangulations.

The observed bearings are usually taken with weight $p = 1$, the weights of the measured sides are computed as explained in Chapter 5.2.

The observation equations for the measured sides read

$$c_{AB}(dx_B - dx_A) + d_{AB}(dy_B - dy_A) + \ell'_{AB} = v'_{AB} \qquad (9.11)$$

where ℓ'_{AB} is the difference computed side minus measured side. If point A or B is fixed, the respective corrections dx and dy are $= 0$.

In the same way, we compute the differences L_{AB} = computed bearing minus observed bearing. For each series, we use as an approximate value of the zero parameter,

$$Z_A = \frac{1}{\nu}\left[L_{AB}\right] \qquad (9.12)$$

where ν = number of bearings in the series observed at point A. Denoting by z_A the correction of this value, the observation equations for the bearings read

$$z_A + a_{AB}(dx_B - dx_A) + b_{AB}(dy_B - dy_A) + \ell_{AB} = v_{AB} \qquad (9.13)$$

where

$$\ell_{AB} = L_{AB} - Z_A \qquad (9.14)$$

The complete scheme of observation equations could be constructed in the following way. Because the parameters z are eliminated first, they should be arranged in the first columns but in each column there would be only one series of figures 1 on the rows of the respective set of bearings. Usually these columns are omitted but two empty rows are left below each series of bearings. Accordingly, the first columns are reserved for the parameters dx and dy only. If both end points A and B of a side are new, the columns dx_A and dx_B have same coefficients a or c with opposite signs on the row of the respective bearing or side observation; the coefficient a with the same sign appears again on the row of the bearing observation from the opposite end of the side. Similar rules apply to coefficients b and d in columns dy_A and dy_B. The last columns are reserved for the absolute terms ℓ or ℓ', for the row sums S and for the weights p.

The elimination of the parameters z can be carried out already during the formation of normal equations. For this purpose, we fill up the first row below each series of bearings computing the sum Σ of figures in each column within each series separately. On the second empty row, we could compute the negative mean values of the same coefficients; in other words, the figures of the row Σ are divided by $-\nu$ where $\nu = 3, 4, \ldots$ is the number of bearings in the set. In our example, Table VI/ 9.6, however, the second row is omitted; instead of it, the weight $-1/\nu$ is written in the column p.

If $\nu = 2$, the two observation equations can be replaced by their difference which represents the observed angle; the weight of such an equation is half of the weight of one bearing.

Each side observation needs a row with coefficients c and d. Usually the next row is reserved for the products pc, pd and $p\ell'$; in our example these extra rows have been omitted too.

When now the normal equations are formed, the parameters z are eliminated by the aid of the additional rows. *E.g.*, when the product sum $[ab] + [pcd]$ is formed, the factor a is taken from row Σ and the factor b from the row of the negative mean.

9.6 *Numerical example*

Fig. 6 shows a part of the second order net in southern Finland. The example is not very typical; the original net is much larger and the observed bearings, because of the failure of the plan made beforehand, have been observed in several different series at some points.

Table I *List of bearings*

No	A → B		Observed	T	T-obs.	l	v
1	29	615	0°00′00″,0	102°36′01″,85	61″,85	+ 2″,29	+1″,59
2		614	57 10 21,9	159 46 21,83	59,93	+ 0,37	+ 77
3		27	91 57 41,6	194 33 38,49	56,89	− 2,67	−2,36
4	31	613	0	33 30 17,97	17,97	− 0,70	−0,51
5		34	31 47 42,9	65 18 02,27	19,37	+ 0,70	+0,50
6	31	614	0	307 37 07,39	07,39	− 0,24	−0,41
7		613	85 53 10,1	33 30 17,97	07,87	+ 0,24	+0,41
8	31	615	0	346 16 28,37	28,37	− 1,46	−1,33
9		614	321 20 36 1	307 37 07,38	31,29	+ 1,46	+1,33
10	32	34	0	136 30 08,47	08,47	− 0,53	−0,48
11		613	50 32 58,3	187 03 08,78	10,48	+ 1,48	+0,50
12		615	132 05 20,6	268 35 28,64	08,04	− 0,96	−0,02
13	613	31	0	213 30 12,12	12,12	− 0,66	−1,39
14		614	48 34 12,3	262 04 25,91	13,61	+ 0,83	+1,33
15		615	82 46 01,4	296 16 13,19	11,79	− 0,99	−0,04
16		32	153 32 58,7	7 03 12,29	13,59	+ 0,81	+0,08
17	614	29	0	339 46 24,66	24,66	+ 0,94	+1,07
18		615	60 36 59,2	40 23 21,97	22,77	− 0,95	−1,08
19	614	615	0	40 23 21,97	21,97	− 0,89	−1,07
20		613	41 41 03,3	82 04 27,04	23,74	+ 0,88	+1,07
21	614	29	0	339 46 24,66	24,66	− 0,38	−0,36
22		31	147 50 38,7	127 37 04,12	25,42	+ 0,38	+0,36
23	615	29	0	282 36 02,64	2,64	+ 0,82	−0,13
24		32	165 59 27,8	88 35 28,79	0,99	− 0,83	+0,12
25	615	32	0	88 35 28,79	28,79	− 1,41	−1,28
26		613	27 40 38,8	116 16 10,40	31,60	+ 1,40	+1,29
27	615	613	0	116 16 10,40	10,40	− 0,16	+0,01
28		31	50 00 10,3	166 16 21,03	10,73	+ 0,17	0
29	615	614	0	220 23 19,08	19,08	+ 0,17	+0,58
30		29	62 12 43,9	282 36 02,64	18,74	− 0,17	−0,58
						[vv] =	25,62

As explanations for the schemes which mostly are selfevident, following statements may be necessary. The approximate coordinates of the new points 613, 614 and 615 have been computed elsewhere and only the accepted results are given in the Table III.

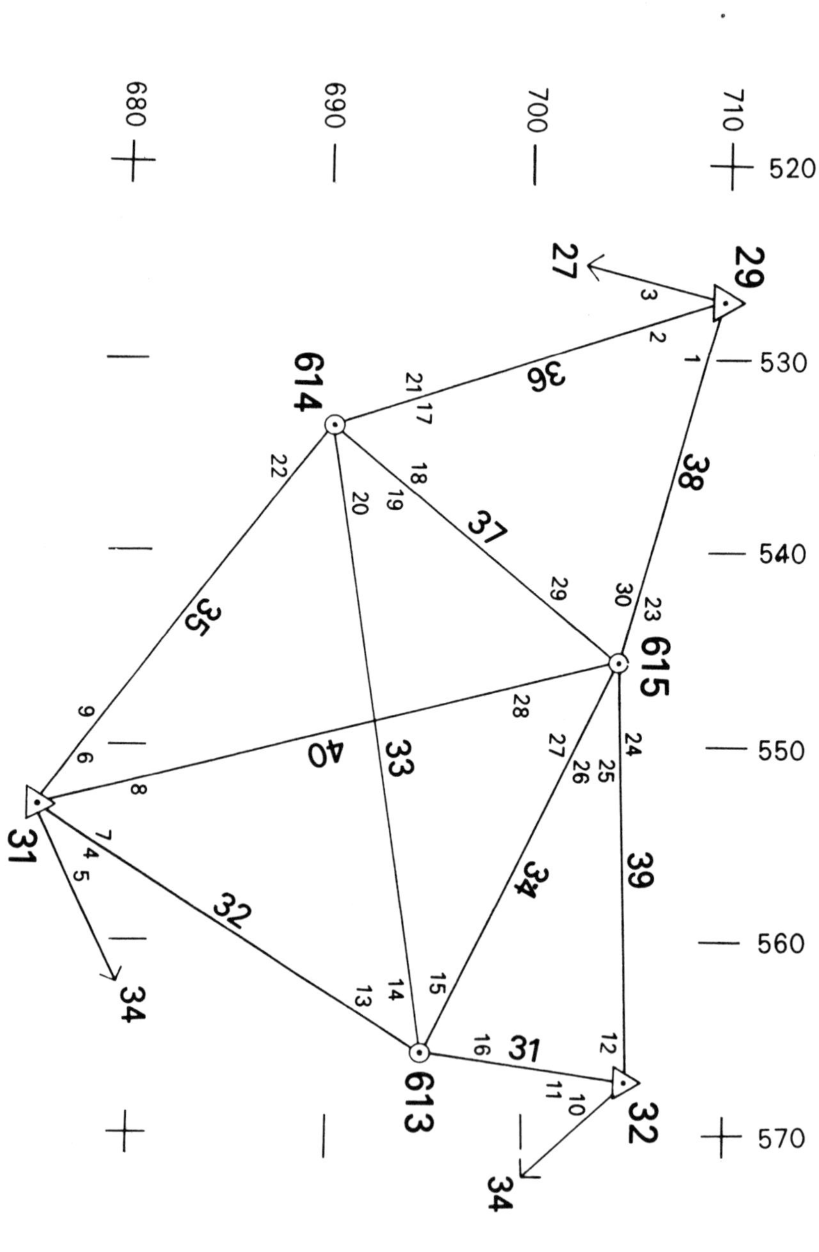

Fig.6. *Sketch of a second order triangulation net.*

Table II *Sides measured by a tellurometer*

No	A → B	Observed	S	S-obs.	p	v
						dm
31	32 613	10 551.511	.516	+0.05	3.1	0.00
32	31 613	23 447.308	.413	+1.05	1.3	+0.51
33	613 614	32 764.886	.889	+0.03	0.7	+0.69
34	613 615	22 468.488	.385	-1.03	1.4	-0.36
35	31 614	24 629.732	.556	-1.76	1.2	-1.50
36	29 614	19 991.341	.268	-0.73	1.6	-0.99
37	614 615	18 988.517	.501	-0.16	1.7	+0.66
38	29 615	19 690.030	.064	+0.34	1.7	+0.37
39	32 615	21 449.876	.889	+0.13	1.5	-0.13
40	31 615	30 363.667	.582	-0.85	0.8	-0.01
		meters	mm	dm	[pvv]	=6.11
						+25.62
						=31.73

Table III *List of coordinates*

P	x	y
27	680 721.88	519 113.59
29	708 901.23	526 433.37
31	675 108.44	552 854.61
32	705 133.04	567 093.75
34	688 790.16	582 600.84
	Approximate	
613	694 660.8	565 798.1
614	690 142.6	533 345.2
615	704 605.7	545 649.5
	Adjusted	
613	694 660.799	565 798.153
614	690 142.621	533 345.183
615	704 605.793	545 649.524
	meters	

The results of the inverse computations are given in Table IV. The observed bearings (Table I) and sides (Table II) do not contain the corrections required by the projection system (Gauss-Krueger); these are given in columns $T - t$ and $S - s$ of Table V where also the coefficients a, b, c and d are listed. The corrections have been

Table IV *Approximate bearings and sides*

A → B		Δx	Δy	t	s
27	29	+28 179.35	+ 7 319.78	14°33'40."20	
29	614	-18 758.62	+ 6 911.83	159 46 23.19	19 991.488
29	615	- 4 295.53	+19 216.13	102 36 02.21	19 690.384
31	34	+13 681.72	+29 746.23	65°18 00.10	
31	613	+19 552.36	+12 943.49	33 30 15.15	23 448.427
31	614	+15 034.16	-19 508.41	307 37 05.63	24 630,125
31	615	+29 497.26	- 7 205.11	346 16 24.61	30 364.485
32	34	-16 342.88	+15 507.09	136 30 11.45	
32	613	-10 472.24	- 1 295.65	187 02 10.54	10 552.086
32	615	- 527.34	-21 444.25	268 35 28.72	21 450.733
613	614	- 4 518.20	-32 452.90	262 04 26.54	32 765.910
613	615	+ 9 944.90	-20 148.60	296 16 11.71	22 469.248
614	615	+14 463.10	+12 304.30	40 23 20.60	18 988.867

Table V *Projection corrections and the coefficients of the observation equations*

A → B		T_1-t_1	T_2-t_2	$S - s$	a	b	c	d
27	29	+ 1."53	-1."71					
29	614	- 1.36	+1.47	-0.220	-0.36	-0.97	-0.94	+0.35
29	615	- 0.36	+0.43	-0.320	-1.02	-0.23	-0.22	+0.98
31	34	+ 2.17						
31	613	+ 2.82	-3.03	-1.014	-0.48	+0.73	+0.83	+0.55
31	614	+ 1.76	-1.51	-0.569	+0.66	+0.51	+0.61	-0.79
31	615	+ 3.76	-3.58	-0.903	+0.16	+0.66	+0.97	-0.24
32	34	- 2.98						
32	613	- 1.76	+1.75	-0.570	+0.24	-1.94	-0.99	-0.12
32	615	- 0.08	+0.07	-0.844	+0.96	-0.02	-0.02	-1.00
613	614	- 0.63	+0.50	-1.021	+0.62	-0.09	-0.14	-0.99
613	615	+ 1.48	-1.31	-0.863	+0.82	+0.41	+0.44	-0.90
614	615	+ 1.37	-1.52	-0.366	-0.70	+0.83	+0.76	+0.65

added to the computed values t and s, the corrected values T and S have been then written in Tables I and II. The weights P of the tellurometer observations have been computed by formula (5.09).

Table VI *Observation equations*

No	613 *x*	613 *y*	614 *x*	614 *y*	615 *x*	615 *y*	*l*	*S*	*p*
1					−1,02	−0,23	+2,29	+1,04	1
2			−0,36	−0,97			+0,37	−0,96	1
3							−2,67	−2,67	1
Σ			−0,36	−0,97	−1,02	−0,23	−1	−2,59	$-\frac{1}{3}$
5−4	+0,48	−0,73					+1,40	+1,15	½
7−6	−0,48	+0,73	−0,66	−0,51			+0,48	−0,44	½
9−8			+0,66	+0,51	−0,16	−0,66	+2,92	+3,27	½
10							−0,53	−0,53	1
11	+0,24	−1,94					+1,48	−0,22	1
12					+0,96	−0,02	−0,96	−0,02	1
Σ	+0,24	−1,94			+0,96	−0,02	−1	−0,77	$-\frac{1}{3}$
13	−0,48	+0,73					−0,66	−0,41	1
14	−0,62	+0,09	+0,62	−0,09			+0,83	+0,83	1
15	−0,82	−0,41			+0,82	+0,41	−0,99	−0,99	1
16	+0,24	−1,94					+0,81	−0,89	1
Σ	−1,68	−1,53	+0,62	−0,09	+0,82	+0,41	−1	−1,46	$-\frac{1}{4}$
18−17			+1,06	+0,14	−0,70	+0,83	−1,89	−0,56	½
20−19	−0,62	+0,09	−0,08	+0,74	+0,70	−0,83	+1,77	+1,77	½
22−21			+1,02	+1,48			+0,76	+3,26	½
24−23					+1,98	+0,21	−1,65	+0,54	½
26−25	−0,82	−0,41			−0,14	+0,43	+2,81	+1,87	½
28−27	+0,82	+0,41			−0,66	+0,25	+0,33	+1,15	½
30−29			−0,70	+0,83	−0,32	−1,06	−0,34	−1,59	½
31	−0,99	−0,12					+0,05	−1,06	3,1
32	+0,83	+0,55					+1,05	+2,43	1,3
33	+0,14	+0,99	−0,14	−0,99			+0,03	+0,03	0,7
34	−0,44	+0,90			+0,44	−0,90	−1,03	−1,03	1,4
35			+0,61	−0,79			−1,76	−1,94	1,2
36			−0,94	+0,35			−0,73	−1,32	1,6
37			−0,76	−0,65	+0,76	+0,65	−0,16	−0,16	1,7
38					−0,22	+0,98	+0,34	+1,10	1,7
39					−0,02	−1,00	+0,13	−0,89	1,5
40					+0,97	−0,24	−0,85	−0,12	0,8

Table VII *Normal equations and the solution by the fill matrix scheme*

	613 *x*	613 *y*	614 *x*	614 *y*	615 *x*	615 *y*	*u*	Σ *S*
613 *x*	6,00	−1,01	+0,06	−0,19	−1,12	+0,57	+1,44	+5,75
y		9,35	−0,05	−0,88	+1,07	−1,23	−5,76	+1,49
614 *x*			4,99	+0,86	−1,57	−0,31	+0,64	+4,62
y				4,97	−1,11	−1,64	+2,55	+4,56
615 *x*					6,62	−0,10	−6,54	−2,75
y						6,80	−1,09	+3,00
							41,41	
j	1	2	3	4	5	6	*c*	*S*
i = 1	6,00	−1,01	+0,06	−0,19	−1,12	+0,57	+1,44	+5,75
2	+0,168	9,16	−0,04	−0,91	+0,88	−1,13	−5,52	+2,44
3	− 10	+0,004	4,99	+0,86	−1,56	−0,32	+0,60	+4,57
4	+ 32	+ 99	−0,172	4,73	−0,79	−1,68	+1,95	+4,21
5	+ 187	− 96	+ 312	+0,167	5,71	−0,27	−5,23	+0,21
6	− 95	+ 123	+ 64	+ 355	+0,046	5,98	−1,42	+4,56
−*d*	−0,240	+0,604	−0,121	−0,413	+0,916	+0,237	9,68	[*pvv*]=
x	−0,008	+0,527	+0,213	−0,174	+0,927	+0,237	9,68	31,73

The normal equations (Table VI) have been solved by the fill matrix scheme (Table VII); another solution by the method of successive approximations is given in Table I/15.1. After the solution, the residuals v are written in Tables I and II. The standard error of unit weight is

$$m = \sqrt{\frac{31.73}{21}} = \sqrt{1.51} = \pm 1''.23.$$

9.7 *Adjustment of a single point*

Schreiber has devised several special rules for the routine connection of single points to a fixed net. Because the usage of these rules requires some experience without which many mistakes can be made, we give them in form of a numerical example. As the material for this example, we shall use the bearing observations of the previous example; there is nothing now concerning the side measurements. Point 613 is supposed to be the only new point; as the fixed points serve 31, 32, 614 and 615 with the definitive coordinates of the preceding adjustment.

Table I summarizes the observations picked up for this example. This time, the absolute terms ℓ are formed from the differences T in the following way: from each difference is subtracted the difference pertaining

- either to the fixed bearing or to the mean of the fixed bearings
- or, at the new point, to the mean of all bearings.

The subtracted value is shown in column "Subtr."

Table I *List of observed bearings*

No	A → B	Observed	T	T-obs.	Subtr.	ℓ	v
4	31 613	0°00'00"0	33°30'17"97	17.97		-1.40	-0.60
5	34	31 47 42.9	65 18 02.27	19.37	19.37	0	+0.59
10	32 34	0	136 30 08.47	08.47		-0.23	-0.31
11	613	50 32 58.3	187 03 08.78	10.48		+1.78	+0.16
12	615	132 05 20.6	268 35 29.53	08.93	08.70	+0.23	+0.15
25	615 32	0	88 35 29.68	29.68	29.68	0	-0.82
26	613	27 40 38.8	116 16 11.26	32.46		+2.78	+0.81
27	615 613	0	116 16 11.26	11.26		+0.22	-0.46
28	31	50 00 10.3	166 16 21.34	11.04	11.04	0	+0.47
13	613 31	0	213 30 12.12	12.12		-0.91	+0.04
14	614	48 34 12.3	262 04 26.06	13.76		+0.73	+0.97
15	615	82 46 01.4	296 16 14.05	12.65		-0.38	-0.78
16	32	153 32 58.7	7 03 12.29	13.59	13.03	+0.56	-0.24

Table II ***Complete observation equations***

No	z_1	z_2	z_3	z_4	z_5	x	y	ℓ	p
4	1					-0.48	+0.73	-1.40	1
5	1							0	1
Σ	2					-0.48	+0.73	-1.40	-1/2
10		1						-0.23	1
11		1				+0.24	-1.94	+1.78	1
12		1						+0.23	1
Σ		3				+0.24	-1.94	+1.78	-1/3
25			1					0	1
26			1			-0.82	-0.41	+2.78	1
Σ			2			-0.82	-0.41	+2.78	-1/2
27				1		-0.82	-0.41	+0.22	1
28				1				0	1
Σ				2		-0.82	-0.41	+0.22	-1/2
13					1	-0.48	+0.73	-0.91	1
14					1	-0.62	+0.09	+0.73	1
15					1	-0.82	-0.41	-0.38	1
16					1	+0.24	-1.94	+0.56	1
Σ					4	-1.68	-1.53	0	-1/4

Table III *Special scheme for the absolute terms*

Point	Forward l	p	Backward l	p	Combined $[pl]$	$[p]$
31	-1.40	1/2	-0.91	1	-1.61	3/2
614			+0.73	1	+0.73	1
615	+2.78	1/2				
	+0.22	1/2	-0.38	1	+1.12	2
32	+1.78	2/3	+0.56	1	+1.75	5/3

Table II gives the complete observation equations with the sum equations for each series of bearings. According to the rules of Schreiber, the observations made at each end of a new side can be combined to one equation; the sum equation is formed only at the new point. The absolute terms of these reduced observation equations are computed by the aid of Table III; they consist of the weighted means $[pl]/[p]$, cfr. Table IV.

Table IV *Reduced observation equations*

P	a	b	l	p	pa	pb
31	-0.48	+0.73	-1.07	3/2	-0.72	+1.10
614	-0.62	+0.09	+0.73	1	-0.62	+0.09
615	-0.82	-0.41	+0.56	2	-1.64	-0.82
32	+0.24	-1.94	+1.05	5/3	+0.40	-3.23
	-1.68	-1.53	0	-1/4	+0.42	+0.38

Table V *Reduced normal equations*

	x	y	u
x	$1.47\ x\ -$	$1.33\ y$	-0.18
y	$-1.33\ x\ +$	$6.83\ y$	-4.96

Solution

$x = +\ 0.95$ dm

$y = +\ 0.91$

As a check, the residuals v have been computed as shown in Table I. From the square sum $[vv] = 4.23$, we obtain the standard error of one observed bearing

$$m = \sqrt{\frac{4.23}{6}} = \sqrt{0.705} = \pm 0''.84$$

Often a mistake is made by computing $[p\ell\ell]$ from the reduced observation equations; then the usual formula (6.13) gives a wrong value for $[vv]$. In this example, $[p\ell\ell]$ must be formed from the original values of ℓ and p as given in the special scheme with the addition of the square sum $[\ell\ell]$ pertaining to the fixed bearings (here rows 10 and 12 of Table I). Then we have $[p\ell\ell] = 8.91$ and formula (6.13) gives

$$[vv] = 8.91 - 0.18 \cdot 0.95 - 4.96 \cdot 0.91 = 4.23$$

as above.

In Chapter 14.1 we shall return to this example for the computation of the error ellipse.

II *ADVANCED COURSE*

10. *Elements of the theory of knowledge*

By the aid of adjustments, we try to make compatible two different kinds of knowledge, one derived from the observations and the other derived from a theory. Too often the former has been considered as "erroneous", the latter as "correct". Therefore, some fundamental ideas of the modern theory of knowledge might be useful.

The *observations*, which word is used here also as meaning experiments and measurements, deal with objects and events which are called "real" if we are convinced that they "exist" even if they are not observed. The entire "reality", however, can not be investigated or exhausted by the aid of the observations, because we are convinced that there are objects and events which have not been or can not be observed.

The *theory* deals with objects and events which are called "concepts" and which exist only in thoughts. The concepts are often "images" of the real objects and events which have been observed earlier but they have been formed by the imagination through abstraction, combination and completion.

We already know very much about the physical phenomena which carry excitations from the objects of observation to our senses; we know something about the physiological phenomena which carry the excitation from the senses to the consciousness, but we know practically nothing about the phenomena which transform the excitations to sensations, memories and concepts. Anyway, there seem to be certain rules governing the internal activity of the consciousness and we are convinced that these rules "exist" even if there were no thinking beings. In other words, in addition to the *physical reality* mentioned above there is another reality which can affect immediately upon our consciousness. To this reality belong, at least, the rules of *logic*.

The communication between two conscious beings is possible by the aid of *symbols*. Each concept should have it's special symbol, *e.g.* a word in a spoken or written language which is mutually understood. The purpose is that when we set a symbol to be observed by another living being, similar mental processes are excited in his consciousness as in our own. In the simplest cases, these processes are uniform with those which are excited by the observation of the original objects and events of the physical reality; already the learning of the languages requires such a correspondence. Therefore, we are used to saying that a word or some other symbol means a real object or event but, to be exact, we can speak and write about concepts only.

The formation of concepts and the learning of the languages seem to involve two different mental operations: *induction* which is based on the observations and *deduction* which complies with the rules of logic. The theory usually is a very highly developed form of deduction which starts from a few statements accepted on the ground of earlier inductions or intuitive hypotheses; the most significant results of a theory are *predictions* about past or future events which then can be compared with the observations. With the increasing knowledge, induction and deduction alternate according to the principle of successive approximations. *E.g.*, the formation of the first concepts is possible only by definitions for which very indefinite concepts already are used.

An observation itself already is an operation for which many earlier observations are essential. In addition, the recording of the results of an observation requires preliminary concepts and symbols. An observed object or event contains an immense number of details from which only a part can be noted and recorded. From many objects and events, we can record details which are similar and others which are different. The similar details are used for the classification which is the first step in the formation of concepts. From the different details, especially those are of special interest which change in permanent combinations or correlations with each other. In this way, all scientific investigation is characterized by one feature: *search of invariances*. We rely on the results of science as far as we believe that which has been invariant in all our observations will remain invariant forever in reality.

The permanent combinations and correlations can often be expressed by mathematical formulas; this is the case especially when the observations consist mainly of measurements. Now the observations pertain to the physical reality, mathematics to the realm of logic. Nowadays we think that mathematics must be independent from the physical world. The mathematical concepts, points, straight lines, an absolute space or time, infinite groups, continuous functions etc., exist only in our imagination; our observations have never found objects or events from the physical reality which correspond exactly to them. However, the predictions which have been made by the aid of the mathematical models upon the physical reality show a far reaching agreement with the observations. For our practical life, the success of the mathematical models has been very beneficial. The extensive and laborious experiments can be replaced by rapid and exact computations. We can even alter the physical reality in order to prevent the unpleasant events or to produce the desired events.

The practical succes of the mathematical models has, in some philosophies of the past centuries, caused the illusion that all observable events could be predicted exactly for the future if the observations could describe completely and exactly the state of the physical universe at the present epoch. The discovery of theoretical laws governing the universe was considered as the ultimate goal of all sciences. Of course, the exact predictions would not be possible in practice because we would never be able to accomplish the description of one momentary state but, nevertheless, the future events were supposed to be predestinate.

First during the present century, the physical theories themselves have assumed aspects which preclude the exact predictions. Instead of them, the new mathematical models give *statistical expectations* only. The complete description of one state of the universe and the exact laws of any logical model can give statements upon the *probability* of various predictions only.

In the theory of adjustments, it is not necessary to go into the extreme causes of the discrepancies between the observed reality and the theoretical predictions. We need a mathematical model for some practical purpose only and we know that it is practically impossible to build a complete theory which would fit the best observations we could make. But we would not even spend time and labor for the best fitting theory which we could find. In adjustments - and perhaps in all sciences - the best theory is only a compromise between two contradictory requirements. It must be *economical*, giving the greatest amount of results for the smallest amount of capital and costs, and, at the same time, *probable*, giving a satisfactory agreement with the observations. The modern base of the adjustment, therefore, is the theory of probability and mathematical statistics.

11. ELEMENTS OF MATHEMATICAL STATISTICS

11.1 *Probability*

Comparison of many observed objects or events can reveal that some of those comprise the same specific detail. The objects or events are said to belong to the same basic group or *population* if they all contain certain specified details.

Let X be a certain detail which does not belong to the specific marks of the population. If we have n observed objects of the same population , X may appear in f objects only. If we have several sets with n objects in each, we may have several different frequency numbers f. If the numbers f or, in a more general case, the relative frequencies f/n do not differ very much from each other, we suppose that there is a constant *probability* p or $p(X)$ which expresses the relative number of objects with detail X in the entire population. Usually we do not know the numerical quantity of p neither the number of the objects in the entire population but if n is a great number, the relative frequency f/n gives a practicable *estimate* for the probability p.

The propability pertains always to the closed interval

$$0 \leq p \leq 1$$

The probability that X does not appear in the population is $1-p(X)$.

Let Y be another detail which has the probability $p(Y)$ in the same population. If X and Y appear in the same observation the occurrence is denoted by XY and it's probability by $p(XY)$. If X or Y or XY appears in one observation the occurrence is denoted by $X + Y$. The *addition of probabilities* can be expressed by formula

$$p(X+Y) = p(X) + p(Y) - p(XY) \qquad (11.01)$$

If X and Y are *mutually excluding events* which can not occur simultaneously, $p(XY) = 0$ and $p(X+Y) = p(X) + p(Y)$.

The observations in which X occurs form another population which is a partial group of the basic group. The event Y which has a certain probability in the second population has in the first population a *conditioned probability*

$$p(Y|X) = \frac{p(XY)}{p(X)} \tag{11.02}$$

The *multiplication of the probabilities* can now be expressed by formula

$$p(XY) = p(X)\cdot p(Y|X) = p(Y)\cdot p(X|Y) \tag{11.03}$$

We set as a definition that Y is *independent* of X if $p(Y|X) = p(Y)$ in the first population. Because then $p(XY) = p(X)\cdot p(Y)$, we can also say that X is independent of Y.

If event Y means that X occured in the nearest antecedent observation, the statement that X is independent of Y can be expressed briefly by saying that two successive observations are independent of each other. In the same way, we could define the statement that all observations are independent of each other.

In our theoretical discussion, we usually assume that the probability is an unknown constant the numerical value of which can be estimated *a posteriori* when a great number of observations has been examined. In certain cases, an estimate can be obtained *a priori* by idealizations and postulations upon the circumstances and appliances.

As a fundamental example, we shall discuss here the *binomial probability*. If the observations are independent of each other and if the probability of an event X is p, then the probability that in n observations X occurs k times is

$$P(k,n) = \binom{n}{k} p^k (1-p)^{n-k} \tag{11.04}$$

where

$$\binom{n}{k} = \frac{n!}{k!\,(n-k)!} \tag{11.05}$$

$$= \frac{n}{1}\,\frac{n-1}{2}\,\frac{n-2}{3}\cdots\frac{n-(k-1)}{k} \tag{11.06}$$

is the *binomial coefficient*, *i.e.* the coefficient of the term a^k in the polynomial development of $(1+a)^n$. For the small values of the exponents, the binomial coefficients can be found from *Pascal's triangle* or from the following table:

Table I *Binomial coefficients* $\binom{n}{k}$

$k=$	0	1	2	3	4	5	6	7	8	9	10	$\Sigma = 2^n$
$n=$ 1	1	1										2
2	1	2	1									4
3	1	3	3	1								8
4	1	4	6	4	1							16
5	1	5	10	10	5	1						32
6	1	6	15	20	15	6	1					64
7	1	7	21	35	35	21	7	1				128
8	1	8	28	56	70	56	28	8	1			256
9	1	9	36	84	126	126	84	36	9	1		512
10	1	10	45	120	210	252	210	120	45	10	1	1024

For large values of n and k, formula (11.05) can be used but the practical computation of the factorial functions can be shortened by approximate expressions like

$$n! = \sqrt{2\pi n}\; n^n\, e^{-n}\left(1 + \frac{1}{12n} + \ldots\right) \qquad (11.07)$$

Often it is advisable to approximate n by an even number and to compute first the greatest coefficient with $k = \frac{n}{2}$

$$\binom{n}{\frac{1}{2}n} = \sqrt{\frac{2}{\pi n}}\left(1 - \frac{1}{4n}\right) \qquad (11.08)$$

E.g., when $n = 10$, $k = 5$, this formula gives 251.91 instead of 252 which is found from the table above. When the greatest coefficient is found, the preceding and succeeding ones can be computed one by one by the aid of the recurrence formula

$$\binom{n}{k+1} = \frac{n-k}{k+1}\binom{n}{k} \qquad (11.09)$$

11.2 *The stochastic variate and it's distribution*

Suppose that each observation consists, in addition to the noting of the specific marks of the population, especially of the measurement of a certain dimension and that the numerical values obtained are scattered over a shorter or longer interval, *range of dispersion*. Suppose further that these values are independent of each other. The dimension thus measured is called a random variable or a *stochastic variate.*

If the measured values are rounded off after a chosen decimal place or if the range of dispersion by other means is divided into a finite number of parts, we can presume that there is a constant probability for each value or part. The collection of these probabilities is called the *distribution.* A sample of values actually measured gives an estimate of the distribution. If the rounded-off values or the chosen parts of the range are represented graphically on the x-axis and if above this axis rectangular areas are drawn which represent the probabilities in a chosen scale, the figure thus obtained is called a *histogram.* It represents a *discrete distribution.*

We can now imagine that the parts are made smaller and smaller until we have an infinite number of infinitely narrow areas. In this case, we have a *continuous distribution* with two probability functions. The *cumulative distribution function*

$$F(x) = p(X \leq x) \qquad (11.10)$$

means the probability that the measured value X is smaller or equal when compared with a chosen value x. The *probability density function*

$$\phi(x) = \frac{dF}{dx} \qquad (11.11)$$

can be used for the computation of the small probability $dF = \phi(x) \cdot dx$ that X lies within a short interval dx.

The density function can be represented by a continuous curve which corresponds to the histogram of a discrete distribution. The area of the figure bounded by the curve and x-axis is always

$$\int_{-\infty}^{+\infty}(x)\ dx = 1 \qquad (11.12)$$

The x-value of the center of gravity of the figure can be obtained from formula

$$x_t = \int_{-\infty}^{+\infty} x\phi(x)dx \qquad (11.13)$$

This value x_t is called the *theoretical mean* or the *expected value:* in the latter case an alternative notation $x_t = E(x)$ is customary. In the case of a discrete distribution, we can write $[p] = 1$ and

$$x_t = [px] \qquad (11.14)$$

If the probabilities are used as weights or if we imagine that the probabilities consists of a mass which is distributed along the x - axis we can consider x_t as a weighted mean or as the static moment of the mass with respect to the origin.

It is advisable to use x_t as a new origin. We denote by

$$\varepsilon = x - x_t \tag{11.15}$$

the deviations of the original stochastic variate from it's theoretical mean. Then we can define the *theoretical variance* for which several different notations are used:

$$E(\varepsilon^2) = \sigma^2 = \sigma_x^2 = \sigma_{xx} = \mu_2 \tag{11.16}$$

In the case of the continuous distribution the definition gives

$$\sigma^2 = \int_{-\infty}^{+\infty} \varepsilon^2 \phi(x) dx \tag{11.17}$$

and in the case of a discrete distribution

$$\sigma^2 = [p\varepsilon\varepsilon] \tag{11.18}$$

If the probability is imagined as a mass the variance is the inertial moment of the mass with respect to the center of gravity. The inertial moment can be computed with respect to the other points too but the formula

$$E(x-x_i)^2 = E(x-x_t)^2 + E(x_t-x_i)^2 \tag{11.19}$$

shows that it is smallest just with respect to the center of gravity.

The numerical value of a variance is always positive. Therefore, we can take a real square root which some times is called *dispersion*. In adjustments, however, we shall deal only with the unbiased estimate of the dispersion and use the customary name *standard error*.

For a detailed analysis of a distribution, still higher moments $\mu_k = E(x-x_t)^k$ could be introduced but they have no application in adjustments, save for the derivation of formulas (11.63) and (11.64).

On the other hand, in adjustments the dispersion is some times replaced by the average deviation

$$\tau = \int_{-\infty}^{+\infty} |\varepsilon| . \phi(x) \, dx \quad \text{or } \tau = [p \cdot |\varepsilon|] \tag{11.20}$$

As examples of theoretical distributions which are supposed to be known a priori we have mentioned in Chapter 2.1 the *even distribution* which is used for the study of the rounding-off errors and the *periodic distribution* which can be studied by harmonic analysis, cfr. Chapter 16.5.

As a further example, formula (11.04) defines the discrete binomial distribution for different values of p and n. In the statistical analysis of infrequent events and accidents, p is usually very small and n very large. If we let n increase and p decrease in such a way that the product $np = a =$ constant, the binomial distribution approaches a limitary case which is continuous. The density function of this *Poisson s distribution* is

$$p_k = \frac{a^k}{k!}e^{-a} \qquad (11.21)$$

Exercises:

1. Four tickets of lottery are bought and the chances are that every fourth ticket will win. What are the probabilities for $k = 0, 1, 2 \ldots$ wins?
2. The same problem when 13 tickets are bought and every 13th wins.
3. A traffic flight is scheduled for each day but the chances are that once a year there will be a revocation. What are the probabilities for k revocations within a year?
4. Compute the probabilities for a Poisson's distribution with $a = 1$.
5. What are the probabilities to receive k aces in a bridge deal?

The answers are given on following page in form of a table:

Exercise	1	2	3	4	5
k = 0	$\frac{81}{256}$ = 0,3164	0,3533	0,3674	0,3679	0,30382
1	$\frac{108}{256}$ = 0,4219	0,3827	0,3684	0,3679	0,43885
2	$\frac{54}{256}$ = 0,2109	0,1913	0,1842	0,1839	0,21349
3	$\frac{12}{276}$ = 0,0469	0,0585	0,0612	0,0613	0,04120
4	$\frac{1}{256}$ = 0,0039	0,0122	0,0152	0,0153	0,00264
5	0	0,0018	0,0030	0,0031	0
6	0	0,0002	0,0005	0,0005	0
7	0	0,0000	0,0001	0,0001	0
Sum	1	1,0000	1,0000	1,0000	1,00000

11.3 *The normal distribution*

Returning to the binomial distribution, we let n increase infinitely but keep $p = \frac{1}{2}$ as a constant. The probabilities $P(k,n)$ of the formula (11.04) can be obtained from the table on p.114 by dividing each number of the row n by $\Sigma = 2^n$. With increasing values of n the rows will become longer but the numbers $P(k,n)$ smaller. From the center $k = n/2$ of each row, the numbers decrease symmetrically in both directions, first slowly, then more rapidly and finally, at the ends of the row, slowly again. We can find two values $k = \frac{n}{2} \pm m_k$ at which the rate of the decrease is greatest from the condition that, close to these points, two consecutive differences are nearly equal.

From formula (11.06) we obtain easily:

$$P(k+1,n) = \frac{p}{1-p}\,\frac{n-k}{k+1}p(k,n)$$

$$P(k-1,n) = \frac{1-p}{p}\,\frac{k}{n-k+1}p(k,n)$$

When $p = 1-p = \frac{1}{2}$, the differences from $P(k,n)$ forwards and backwards are

$$\delta_1 = \frac{n-2k-1}{k+1}P(k,n) \quad \text{and} \quad \delta_2 = \frac{n-2k+1}{n-k+1}P(k,n)$$

These differences are equal when

$$k = \frac{n}{2} \pm \frac{\sqrt{n+2}}{2}$$

When n is a large number, we can put approximately $m_k = \sqrt{n}/2$.

Let us now introduce a new system of coordinates x, y in such a way that $x = 0$ when $k = n/2$, $x = 1$ when $k = \frac{n}{2}+m_k$ and that the area bounded by the histogram remains equal to 1. Then

$$x = \frac{2k-n}{\sqrt{n}} \qquad y = \frac{1}{2}\sqrt{n}\,P(k,n) \tag{11.22}$$

When k increases one unit, x increases by a small amount

$$dx = \frac{2}{\sqrt{n}} \tag{11.23}$$

When n is a large number we can write

$$\frac{dy}{y} = \frac{\delta}{P(k,n)} = -\frac{\sqrt{n}\,x}{\frac{n}{2} \pm \frac{x}{2}\sqrt{n}}$$

We let now n increase infinitely but consider only finite values of x. Then the second term in the denominator vanishes and we obtain

$$\frac{1}{y}\frac{dy}{dx} = -x \tag{11.24}$$

The general solution of this differential equation is

$$\ln y = \ln c - \frac{x^2}{2} \tag{11.25}$$

or

$$y = c\exp(-\frac{x^2}{2}) \tag{11.26}$$

The constant of integration can be computed from the area bounded by the curve representing the function (11.26):

$$\int_{-\infty}^{+\infty} y\,dx = 1 \tag{11.27}$$

or $cI = 1$ where

$$I = \int_{-\infty}^{+\infty} \exp(-\frac{x^2}{2})dx = \int_{-\infty}^{+\infty} \exp(-\frac{z^2}{2})dz$$

The definite integral I can be evaluated in such a way that first the value of

$$I^2 = \int_{-\infty}^{+\infty}\int_{-\infty}^{\infty} \exp(-\frac{x^2+z^2}{2})dx\,dz$$

is evaluated by substitutions $x^2+z^2 = r^2$ and $dx\,dz = r\,dr\,du$:

$$I^2 = \int_0^{2\pi}\int_0^{\infty} \exp(-\frac{r^2}{2})r\quad dr\,du$$

which gives easily $I^2 = 2\pi$ and hence $c = 1/\sqrt{2\pi}$. Then (11.26) gives

$$y = \frac{1}{\sqrt{2\pi}}\exp(-\frac{x^2}{2}) \qquad (11.28)$$

or, in form of a computation formula,

$$\log\ y = -0{,}217147\ x^2 - 0{,}399090 \qquad (11.29)$$

This is the probability density function of a continuous distribution which is the limiting case $n = \infty$ of the binomial distribution. It is called the *standardized normal distribution*. Table I gives the numerical values of both the density function y and the cumulative function F. Fig. 7 shows the probability values of the binomial distribution for cases $n = 1, 2, \ldots 10$ in the coordinate system defined by (11.22) and, for the limit case of the normal distribution, the corresponding continuous curve, called *Gaussian error curve*.

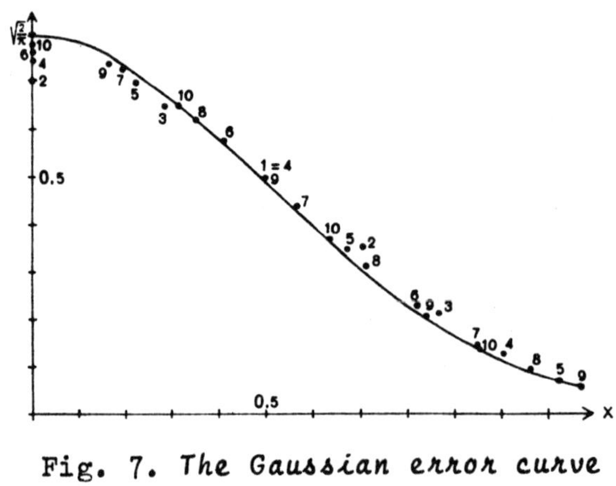

Fig. 7. *The Gaussian error curve*

Table I *The standardized normal distribution*

x	y	F	x	y	F
0.0	0.3989	0.5000	2.0	0.0540	0.9772
0.1	3970	5398	2.1	440	9821
0.2	3910	5793	2.2	355	9861
0.3	3814	6179	2.3	283	9893
0.4	3683	6554	2.4	224	9918
0.5	3521	6915	2.5	175	9938
0.6	3332	7257	2.6	136	9953
0.7	3123	7580	2.7	104	9965
0.8	2897	7881	2.8	79	9974
0.9	2661	8159	2.9	60	9981
1.0	2420	8413	3.0	44	9987
1.1	2179	8643	3.1	33	9990
1.2	1942	8849	3.2	24	9993
1.3	1714	9032	3.3	17	9995
1.4	1497	9192	3.4	12	9997
1.5	1295	9332	3.5	9	9998
1.6	1109	9452	3.6	6	9998
1.7	940	9554	3.7	4	9999
1.8	790	9641	3.8	3	9999
1.9	656	9713	3.9	2	1.0000
2.0	540	9772	4.0	1	1.0000

The table is computed, with rounded-off figures, up to $x = 4.0$ but actually there is no limit. Any large value of the deviation x is possible but extremely improbable. The probability that x is greater than 3 is $1 - 2\,F(3.0) = 0.0026$ or, more exactly, 1/370 which means that in a sample of 370 items there is, in average, one x greater than 3. In practice, when the samples are smaller, such an exceptional value is suspected to be a "blunder", an item which does not belong to the population.

The variance of a stochastic variate with standardized normal distribution is according to (11.17)

$$\sigma_x^{\,2} = \int_{-\infty}^{+\infty} y x^2 dx = 2 \int_{\infty}^{\infty} y x^2 dx$$

According to (11.24), this gives

$$\sigma_x^{\,2} = -2 \int_{y_0}^{0} x\, dy$$

Integration by parts gives

$$\int x\, dy = xy - \int y\, dx$$

Because the product xy vanishes at both boundaries, (11.27) gives

$$\sigma_x^2 = 1$$

In other words, the unit which was originally chosen to correspond to the inflection point of the curve means, at the same time, the dispersion σ_x.

Any random variable which has a standardized normal distribution is called *standardized normal stochastic variate.* Generally, a *normal stochastic variate* u can be standardized by the transformation formula.

$$x = \frac{u - u_t}{\sigma_u} \qquad (11.30)$$

From (11.28) we can conclude that the general *density function of the normal distribution* is

$$\phi(u) = \frac{1}{\sigma_u\sqrt{2\pi}}\exp\left\{-\frac{1}{2}\left(\frac{u - u_t}{\sigma_u}\right)^2\right\} \qquad (11.31)$$

The average deviation of a standardized normal stochastic variate is according to (11.20)

$$\tau = 2\int_0^\infty yx\ dx = -2\int_{y_0}^0 dy = \frac{1}{\sqrt{2\pi}} = 0.79788$$

or about 4/5 from the dispersion, cfr. (2.11).

The *probable deviation* is defined by the equation

$$F(x) - F(-x) = \frac{1}{2}$$

A numerical integration by successive approximations gives the solution

$$x = r = 0.67449$$

or about 2/3 from the dispersion, cfr. (2.12)

The fourth moment with respect to the center of gravity of the standardized normal variable is

$$\mu_4 = \int_{-\infty}^{+\infty} yx^4 dx = -2\int_{y_0}^0 x^3 dy$$

Integration by parts gives $\mu_4 = 3$.

For a general normal variable we could get in the same way

$$\mu_4 = 3\ \sigma_u^4 \qquad (11.32)$$

The normal distribution has some theoretical specialities from which we mention two theorems without proof.

1. If several mutually independent stochastic variates have a normal distribution, their sum has a normal distribution too.

2. The sum of n mutually independent stochastic variates with arbitary distributions approaches, when n increases, to a normal distribution, provided that no one of the original variables is dominating.

The latter theorem explains the fact that the distribution of empirical stochastic variates very often is nearly normal. This fact gives a theoretical argument in favor of the method of least squares. For if we suppose that the residuals after an adjustment all represent normal stochastic variates, then the probability y_i that one residual obtains the value v_i can be computed according to (11.31)

$$\ln y_i = \text{const.} - \frac{v_i^2}{2\sigma_i^2} \tag{11.33}$$

The probability that all residuals $v_1, v_2, \ldots v_n$ occur simultaneously is according to (11.03) $y_1\, y_2, \ldots\, y_n$. This probability and it's logarithm are greatest when $\left[\left(\frac{y}{\sigma}\right)^2\right]$ has it's minimum value. When the weight p_i is inversely proportional to the variance σ_i^2, the said condition can be written in form $[pvv]$ = minimum. We conclude, that an *adjustment by the method of least squares has given the most probable values to the observed quantities if the residuals have a normal distribution.* Note, however, the remarks given on p.13.

11.4 *Examination of an empirical distribution*

Suppose that we have large number N of observed quantities pertaining to the same population. For a statistical analysis, the range of dispersion is divided into equal intervals the number of which is of order $\sqrt{N}$; then the frequencies are, in average, of order $\sqrt{N}$ too.

Each interval is plotted on the x - axis, each frequency is represented by a rectangular area the base of which is formed by the interval. This histogram gives the first illustration about the distribution. If it has a symmetrical appearance, the center of that interval or that boundary value between two intervals which is most close to the center of symmetry is taken as a rough mean value x_m. The frequencies of each pair of symmetrical intervals are added up and the frequency of the possible odd interval at the center multiplied by 2. These new frequencies f give a halved histogram which usually is more regular and which easily can be replaced by a continuous curve. The difference $x_d - x_m$ where x_d corresponds to the inflexion point of the curve gives a rough estimate of the dispersion.

This *graphical examination* can be accomplished also in an other way. Denoting the deviation $x_i - x_m$ by ε, the values of ε^2 are marked on the horisontal axis. The logarithms of the frequencies f are measured in the vertical direction. The points thus plotted lie all very close to a straight line if the distribution is normal; cfr. fig.8. By the aid of this line, we can read the value ε_a^2 corresponding to the intersection with the horisontal axis and the value $\log f_o$ corresponding to the intersection with the vertical axis. The variance can now be estimated by

$$m^2 = 0.217 \frac{\varepsilon_a^2}{\log f_o} \tag{11.34}$$

A *numerical analysis* begins with the computation of the mean value and of the variance by the aid of formulas (11.14) and (11.18). In addition, formula (11.20) gives the average deviation and a simple counting the middlemost absolute deviation. Then the formulas (2.11) and (2.12) can be tried; if the agreement is satisfactory, we can conclude that the distribution is fairly normal.

The greatest absolute deviation should not be more than $k=3$ times the dispersion. An exact discussion would show, however, that this value is not constant but depends on N according to the following table which has been suggested by Chauvenet:

N	10	20	30	50	100	500
k	2.0	2.2	2.3	2.5	2.8	3.3

A deviation which is greater than k times the dispersion could be interpreted as a blunder and the due observation should be rejected or renewed. A more modern method for the computation of a "confidence interval" is given in Chapter 11.7.

The *detailed numerical examination* of the distribution could be accomplished by the comparison of the empirical frequencies f with the theoretical frequencies $N\phi$ where ϕ can be computed by formula (11.31) or using Table I/11.3. As a sufficient approximation, we can replace the normal distribution by a binomial distribution. Let n be an integer as close to $2m_i^2$ as possible. The theoretical frequency pertaining to the interval at the center of symmetry is

$$\phi_o = 0.399 \frac{N}{m_i} \qquad (11.35)$$

The other frequencies can be obtained from formulas

$$\phi_1 = \phi_o \frac{n}{n+1}, \quad \phi_2 = \phi_1 \frac{n-1}{n+2}, \quad \phi_3 = \phi_2 \frac{n-2}{n+3}, \quad \text{etc.} \qquad (11.36)$$

As an example, we shall examine the misclosures of the first order triangles measured by the Finnish Geodetic Institute in 1920-1947. Originally, these misclosures have been computed and published with an accuracy of 0.01 sexagesimal seconds but here they have been grouped with intervals of 0".15.

Distribution of misclosures in a first order triangulation

i	xi	f^-	f^+	f	if^-	if^+	fi^2	f_t	$f^- - f_t$	$f^+ - f_t$	$f - f_t$
0	0",00			29			0	26,9			+2
1	0,15	23	37	60	23	37	60	26,1	−3	+11	+8
2	0,30	26	14	40	52	28	160	23,9	+2	−10	−8
3	0,45	20	24	44	60	72	396	20,6	−1	+3	+3
4	0,60	17	14	31	68	56	496	16,7	0	−3	−2
5	0,75	10	16	26	50	80	650	12,7	−3	+3	+1
6	0,90	8	8	16	48	48	576	9,1	−1	−1	−2
7	1,05	7	5	12	49	35	588	6,2	+1	−1	0
8	1,20	3	4	7	24	32	448	3,9	−1	+0	−1
9	1,35	1	2	3	9	18	243	2,3	−1	−0	−2
10	1,50		2	2		20	200	1,3	−1	+1	−1
11	1,65		1	1		11	121	0,7	−1	0	0
12	1,80		1	1		12	144	0,3	−0	+1	0
13	1,95		1	1		13	169	0,2	−0	+1	+1
16	2,40		1	1		16	256	0,0	0	+1	+1
Sum		115	130	274	383	478	4507	(275)	−8	+7	0

There are more positive than negative misclosures but the relative difference

$$\frac{[\delta^+] - [\delta^-]}{[\delta]} = \frac{15}{274}$$

is smaller than what in average could be expected. More striking is the fact that the six greatest misclosures are all positive but no plausible hypotheses for the explanation of this fact has yet been suggested.

As to the arithmetic mean of misclosures, we compute

$$i_m = \frac{478-383}{274} = +0.35, \quad x_m = 0''15 \cdot i_m = +0''052$$

The variance of i is 4507/274 = 16.4 from which we compute the dispersions

$$m_i = \pm 4.06, \quad m_x = \pm 0''608$$

The standard error of the arithmetic mean is $0''608/\sqrt{274} = 0''037$. The actual arithmetic mean 0.052 is 1.4 times as great as that, the probability of which is, according to Table I/11.3, 2(1-0.92) =0.16 or once in six similar triangulations.

The greatest misclosure 2"40 is four times as great as the standard deviation. Such an event could be expected once in 10000 triangles. All these facts point to some unknown disturbance.

The middlemost misclosure in our table is 0"45, in the original compilation 0"42. According to a normal distribution, it should be $0.6745 \cdot 0''608 = 0''41$. The average deviations are

$$\tau_i = \frac{383+478}{274} = 3.14, \quad \tau_x = 0''47$$

According to a normal distribution, it should be $0.7979 \cdot 0''608 = 0''485$ The suspected disturbance does not show up in these comparisons.

A graphical representation of the distribution is given in Fig.8 From this figure we obtain rough estimates

$$\ln \delta_0 = 4.0, \quad i_a^{\,2} = 131$$

and according to formula (11.34),

$$m_i^{\,2} = \frac{131}{8.0} = 16.4.$$

A detailed numerical examination is accomplished by formulas (11.35) and (11.36) using the value $n = 33$. The results have been added to the table given above. There are only two marked deviations between the empirical and theoretical frequencies. As they have opposite signs and appear in neighboring groups, they would vanish from broader groups.

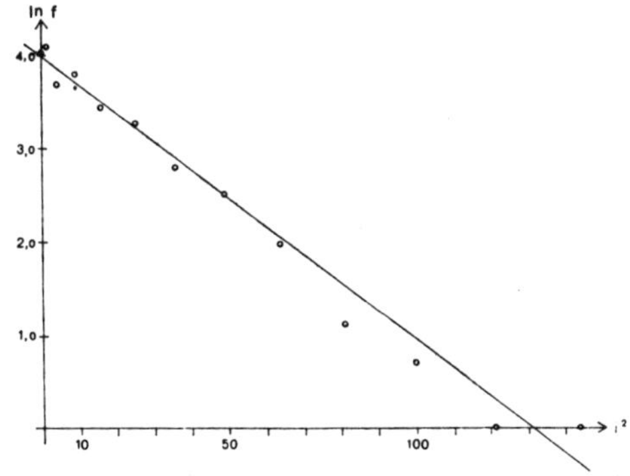

Fig. 8. *The logarithm of an error function*

11.5. *Two-dimensional stochastic variates*

Suppose that each experiment or each measuring process gives simultaneously discrete values for two different quantities x and y. The results of a great number of experiments can be collected into a table where each possible value of x is represented by a row, each value of y by a column. Into each square of the table can be written the frequency $f(x,y)$ which shows how many times the corresponding couple of values x,y has occurred. The row sums and the column sums show the frequencies of each x and y separately. When all frequencies are divided by the total number of experiments, the relative frequencies are obtained. We suppose that these relative frequencies are more or less stable when the set of experiments is repeated several times.

Now we presume that there are *a priori* certain constant probabilities p_{xy} for the occurrence of each couple of values x,y in one experiment. If the table of frequencies is replaced by a similar table of numbers p_{xy}, the row sums and the column sums signify the *border probabilities*

$$p_x = \sum_y p_{xy} \qquad \text{and} \qquad p_y = \sum_x p_{xy} \tag{11.37}$$

If x and y are independent of each other,

$$p_{xy} = p_x p_y \tag{11.38}$$

In theory, we can imagine that the rows and the columns can be divided into smaller and smaller parts until, in the limiting case, we obtain a continuous distribution instead of the discrete distribution. Instead of the discrete numbers p_{xy}, we have a density function of two variables: $\phi(x,y)$. We can also define the border densities

$$\phi_1(x) = \int_{-\infty}^{+\infty} \phi(x,y)\, dy$$
$$\phi_2(y) = \int_{-\infty}^{+\infty} \phi(x,y)\, dx \qquad (11.39)$$

and the cumulative function

$$F\,(x,y) = \int_{-\infty}^{x} \int_{-\infty}^{y} \phi(x,y)\, dy\, dx \qquad (11.40)$$

The conditional densities are defined by

$$\phi(y|x) = \frac{\phi(x,y)}{\phi_1\,(x)}$$
$$\phi(x|y) = \frac{\phi(x,y)}{\phi_2\,(y)} \qquad (11.41)$$

If x and y are independent of each other,

$$\phi(x,y) = \phi_1(x) \cdot \phi_2(y) \qquad (11.42)$$

Usually, however, x and y are correlated. The extreme degree of correlation is a functional dependence $y = y(x)$ in which case the graphical representation of the distribution is a curve.

Both x and y can be considered also as a onedimensional stochastic variate with a theoretical mean, variance and higher moments:

$$x_t = E(x) = \int_{-\infty}^{+\infty} x\phi_1(x)\, dx = \int_{-\infty}^{+\infty}\int_{-\infty}^{+\infty} x\phi(x,y)\, dx\, dy \qquad (11.43)$$
$$y_t = E(y) = \int_{-\infty}^{+\infty} y\phi_2(y)\, dy = \int_{-\infty}^{+\infty}\int_{-\infty}^{+\infty} y\phi(x,y)\, dx\, dy$$
$$\sigma_x^2 = E(x-x_t)^2 = \int_{-\infty}^{+\infty}\int_{-\infty}^{+\infty} (x-x_t)^2\phi(x,y)\, dx\, dy \qquad (11.44)$$
$$\sigma_y^2 = E(y-y_t)^2 = \int_{-\infty}^{+\infty}\int_{-\infty}^{+\infty} (y-y_t)^2\phi(x,y)\, dx\, dy$$

But now we must define a new important quantity, the *covariance:*

$$\sigma_{xy} = E(x-x_t)(y-y_t) = \int_{-\infty}^{+\infty}\int_{-\infty}^{+\infty}(x-x_t)(y-y_t)\phi(x,y)\,dx\,dy \quad (11.45)$$

which often is replaced by the correlation coefficient

$$\rho_{xy} = \frac{\sigma_{xy}}{\sigma_x\sigma_y} \quad (11.46)$$

An empirical correlation is often examined graphically by the aid of the frequency tables in such a way that the entire region of dispersion is halved by a continuous line in the longitudinal direction. In fact, two *regression curves* can be found if the center of gravity is marked up first for each row, then for each column. The mathematical formulas for these regression curves read

$$y = E(y|x) \quad \text{and} \quad x = E(x|y) \quad (11.47)$$

In *regression analysis,* the regression curves are often replaced by straight regression lines which are determined by the method of least squares according to formula (16.06):

$$y = y_t + \frac{\sigma_{xy}}{\sigma_x^2}(x-x_t) \quad (11.48)$$

$$x = x_t + \frac{\sigma_{xy}}{\sigma_y^2}(y-y_t)$$

The two-dimensions stochastic variates can be *standardized* by the aid of mathematical functions. Suppose first that x and y have been replaced by a couple of new parameters u and v:

$$u = u(x,y) \quad (11.49)$$

$$v = v(x,y)$$

These functions define a map representation from the x,y-plane on a u,v-plane when the variables are used as rectangular coordinates. If we take from the xy-plane a small rectangle with sides dx and dy, we have on the uv-plane a small parallelogram with area $D\,dx\,dy$. The coefficient

$$D = \frac{\partial u}{\partial x}\frac{\partial v}{\partial y} - \frac{\partial u}{\partial y}\frac{\partial v}{\partial x} \quad (11.50)$$

is called the *functional determinant.*

If $D \neq 0$, the functions (11.49) can be inverted in form

$$x = x(u,v)$$
$$y = y(u,v) \qquad (11.51)$$

If the density function of the two-dimensional stochastic variate x,y is $\phi(x,y)$, that of the new variate u,y is

$$\Psi(u,v) = D\phi\left\{x(u,v),y(u,v)\right\} \qquad (11.52)$$

In the practice the functions (11.49) are usually replaced by linear approximations

$$u = a_1x + b_1y + c_1$$
$$v = a_2x + b_2y + c_2 \qquad (11.53)$$

Then we have $D = a_1b_2 - b_1a_2$. We require now that x and y are *uncorrelated standardized normal variables*, in other words; $x_t = y_t = 0$, $\sigma_x = \sigma_y = 1$ and $\sigma_{xy} = 0$. Then we have

$$u_t = c_1 \quad \text{and} \quad v_t = c_2 \qquad (11.54)$$

$$\sigma_u^2 = a_1^2 + b_1^2 \quad \text{and} \quad \sigma_v^2 = a_2^2 + b_2^2 \qquad (11.55)$$

$$\sigma_{uv} = a_1a_2 + b_1b_2 \qquad (11.56)$$

According to (11.03) and (11.28), the density function of x,y is

$$\phi(x,y) = \frac{1}{2\pi}\exp\left(-\frac{x^2+y^2}{2}\right) \qquad (11.57)$$

According to (11.52), we obtain now the density function of u,v:

$$\Psi(u,v) = \frac{D}{2\pi}\exp\left(-\frac{F}{2}\right) \qquad (11.58)$$

where

$$F = x^2y^2 = \frac{1}{D^2}\left\{\sigma_v^2(u-u_t)^2 + \sigma_u^2(v-v_t)^2 - 2\sigma_{uv}(u-u_t)(v-v_t)\right\}$$

Formula (11.58) represents the *two-dimensional normal distribution* in a general case.

All the discussions of this chapter can be generalized for three- and multidimensional stochastic variates. If the number of dimensions or components is ν, that of the covariances which must be taken for pairs in all combinations is $\frac{1}{2}\nu(\nu-1)$.

11.6. *Samples and estimates*

The characteristics of a distribution

theoretical mean $x_t = E(x)$ (11.13)

theoretical variance $\sigma_x^2 = E(x-x_t)^2$

higher moments $\mu_k(x) = E(x-x_t)^k$ (11.16)

are usually unknown constants, the determination of which would require that all possible values of x with due frequencies be known. In practice, we have only a finite number n of values x_i ($i = 1, 2, \ldots n$). A random set of such values is called a *sample*. By the aid of a sample we can compute *estimates* for the unknown constants:

mean value $$x_m = \frac{[x]}{n} \qquad (11.59)$$

variance $$m_x^2 = \frac{[(x-x_m)^2]}{n-1} = \frac{[vv]}{n-1} \qquad (11.60)$$

These estimates are also stochastic variates for new samples would give slightly different estimates. Each of them has it's own distribution which, however, can be theoretically determined from the distribution of x itself.

An estimate is *unbiased* if it's theoretical mean is equal to the characteristic to be estimated. The estimates x_m and m_x^2 given above meet this requirement; for the former it is evident, for the latter the proof was given in Chapter 4.1.

We shall consider now the theoretical variance of an estimate. In Chapter 3.2, we have already shown that the variance of the mean value is

$$E(x_m-x_t)^2 = \frac{\sigma_x^2}{n} \qquad (11.61)$$

The theoretical variance of the sample variance (11.60) requires a longer explanation. It is defined by formula

$$E(m_x^2-\sigma_x^2)^2 = E\left(\frac{[vv]}{n-1} - \sigma_x^2\right)^2 = E\left(\frac{[vv]}{n-1}\right)^2 - \sigma_x^4 \qquad (11.62)$$

Formula (4.10) gives first

$$[vv] = [\varepsilon\varepsilon] - \frac{[\varepsilon]^2}{n}$$

from which follows

$$[vv]^2 = [\varepsilon^2]^2 - \frac{2}{n}[\varepsilon^2][\varepsilon]^2 + \frac{[\varepsilon]^4}{n^2}$$

By the aid of equations

$$[\varepsilon^2]^2 = [\varepsilon^4] + 2[\varepsilon_i^2\varepsilon_j^2]$$

$$\bullet [\varepsilon^2][\varepsilon]^2 = [\varepsilon^4] + 2[\varepsilon_i^2\varepsilon_j^2] + 2[\varepsilon^2][\varepsilon_i\varepsilon_j]$$

$$[\varepsilon]^4 = [\varepsilon^4] + 6[\varepsilon_i^2\varepsilon_j^2] + 4[\varepsilon_i^3\varepsilon_j] + 12[\varepsilon_i^2\varepsilon_j\varepsilon_k] + 24[\varepsilon_i\varepsilon_j\varepsilon_k\varepsilon_l]$$

we can compute the expected value (11.62) because

$$E([\varepsilon^4]) = n\mu_4$$

$$E([\varepsilon_i^2\varepsilon_j^2]) = \tfrac{1}{2}n(n-1)\sigma_x^4$$

$$E([\varepsilon_i\varepsilon_j]) = E([\varepsilon_i^3\varepsilon_j]) = E([\varepsilon_i^2\varepsilon_j\varepsilon_k]) = E([\varepsilon_i\varepsilon_j\varepsilon_k\varepsilon_l]) = 0$$

Finally we obtain

$$E(m_x^2 - \sigma_x^2)^2 = \frac{1}{n}\mu_4 - \frac{n-3}{n(n-1)}\sigma_x^4$$

We suppose now that x has a normal distribution. According to (11.32), we have then $\mu_4 = 3\sigma_x^4$ and the expression above becomes

$$E(m_x^2 - \sigma_x^2)^2 = \frac{2}{n-1}\sigma_x^4$$

The square root of this expression means the dispersion of the sample variance. It has the same dimension as m_x^2 but, in practice, it is much smaller. In the usual way, we write the dispersion as a differential variation with double sign. For the variance we obtain

$$m_x^2 \pm \sigma_x^2\sqrt{\frac{2}{n-1}} \approx m_x^2(1 \pm \sqrt{\frac{2}{n-1}})$$

and for the dispersion

$$m_x(1 \pm \sqrt{\frac{1}{2(n-1)}}) = m_x \pm \frac{m_x}{\sqrt{2(n-1)}}$$

This formula shows that the *standard error of the standard error* m_x can be estimated by

$$m_{m_x} = \frac{m_x}{\sqrt{2(n-1)}} \qquad (11.63)$$

After an adjustment, the standard error of the weight unit is estimated by $m = \sqrt{\frac{[vv]}{n-u}}$, cfr. (13.58). Without further proofs, we give for it the corresponding formula

$$m_m = \frac{m}{\sqrt{2(n-u)}} \tag{11.64}$$

11.7. *Confidence intervals*

If u is a normal variable with theoretical mean u_t and dispersion σ, then the mean value u_m of a sample with n observed values of u is also a normal variable with theoretical mean u_t and dispersion $\sigma/\sqrt{n}$. Then $\sqrt{n}(u_m - u_t)/\sigma$ is a standardized normal variable. The table of p.121 shows that the value of such a variable is between limits -1 and +1 with probability 2(0.8413 - 0.5000) = 0.6826. With the same probability or, in other words, on the *confidence level* of 68 %, u_t is between limits

$$u_m - \frac{\sigma}{\sqrt{n}} < u_t < u_m + \frac{\sigma}{\sqrt{n}} \tag{11.65}$$

In statistical investigations, we usually want a greater confidence to the statement that a value is between two given limits or within a *confidence interval* which, of course, then must be greater too. The level of confidence which usually is expressed in percents is denoted here by $1 - \varepsilon$ where ε is a small fraction. From the table of the standardized normal distribution we shall find such value of x that the corresponding probability is $p = 1 - F = \varepsilon/2$. The results are given in Table I:

Table I *Standardized normal distribution*

p	0.1	0.05	0.025	0.02	0.01	0.005	0.0005
x	1.28	1.64	1.96	2.05	2.33	2.58	3.29

The confidence interval is then

$$u_m - \frac{x\sigma}{\sqrt{n}} < u_t < u_m - \frac{x\sigma}{\sqrt{n}} \tag{11.66}$$

As an example, in Chapter 11.4 we had x_m = 0"052 and m = 0"037. On the confidence level 95 %, the confidence interval is 0"052 ± 1.96 · 0"037 or −0"021 < x_t < +0"124. Because the ideal expected value x_t = 0, is within this interval we may state that the deviation $x_m - x_t$ is not *significant.*

Even if u has a normal distribution, the distribution of a sample variance m_u^2 is not normal. Suppose that the residuals $v_i = u_m - u_i$ have a normal distribution, then the square sum

$$s = v_i^2 + v_2^2 + \ldots + v_n^2 \tag{11.67}$$

is a stochastic variate which has so called *chi-square* or χ^2 *-distribution.* The density function of χ^2-distribution is

$$k(s) = C_n \sqrt{s^{n-2} e^{-s}} \tag{11.68}$$

where the coefficient C_n can be computed by the aid of Euler's gammafunctions:

$$\frac{1}{C_n} = \sqrt{2^n \Gamma(\tfrac{n}{2})} \tag{11.69}$$

$$\Gamma(z) = \int_0^\infty t^{z-1} e^{-t} dt \tag{11.70}$$

When z is an integer, we can write $\Gamma(z) = (z-1)!$, which can be used also for all great values of z in the connection of the approximate computation formula (11.07).

For practical computations, we should have first a table from which we could find the cumulative functions p of the χ^2 -distribution for several parameters n and for different values of s. Table II below shows the values of s which correspond to the values $p_1 = 0.02$ and $p_2 = 0.98$. The confidence interval on the level $p_2 - p_1$ is then

$$m\sqrt{\frac{n}{s_1}} < \sigma < m\sqrt{\frac{n}{s_2}} \tag{11.71}$$

When $n > 20$, we can use Table I for the probability p, find x and compute the approximate values

$$\sqrt{s} = \sqrt{n - \frac{1}{2}} \pm \frac{x}{\sqrt{2}} \tag{11.72}$$

Table II *Chi-square-distribution*

n	1	2	3	4	5	6	7	8	9	10
s_1	5.41	7.82	9.84	11.67	13.39	15.03	16.62	18.17	19.68	21.16
s_2	0.00	0.04	0.18	0.43	0.75	1.13	1.56	2.03	2.53	3.06
n	11	12	13	14	15	16	17	18	19	20
s_1	22.62	24.05	25.47	26.87	28.26	29.63	31.00	32.35	33.69	35.02
s_2	3.61	4.18	4.76	5.37	5.98	6.61	7.26	7.91	8.57	9.24

As an example, in Chapter 11.4 we had $m_x = 0\overset{''}{.}608$ and $n = 274$. From Table I we obtain for $p = 0.025$ the value $x =$ 1.96 which gives from (11.72) $s_1 = 321$, $s_2 = 230$. The confidence interval on level 95 % is, according to (11.71), $0\overset{''}{.}562 < \sigma <$ $0\overset{''}{.}663$. Formula (11.63) for the standard error of a standard error gives $\sigma = 0\overset{''}{.}608 \pm 0\overset{''}{.}026$.

The confidence interval is a refinement of the old concepts "range of dispersion" or "maximum error" because it indicates an interval within which a stochastic variate lies with a great and exactly declared probability when the variate has a normal distribution.

12. *MATRICES*

12.1 *Notations*

The extensive numerical computations which are necessary for the adjustments by the method of least squares are usually carried out by the aid of tables with rows and columns as we already have seen in Chapter 9. The *matrix* is nothing else than such a table; the numbers written into the matrix are called *elements*. When a whole matrix is denoted by a simple symbol and when some rules for the computational operations have been accepted, a great number of similar formulas for a great number elements can be given in a very short form like a command to soldiers which have been arranged in rows and columns.

The rectangular or two-dimensional matrices are often denoted by boldfaced capital letters. In this book, however, we shall use Roman type letters. When it is necessary to indicate how many rows and columns a matrix contains, these *dimensions* of the matrix are written below the main symbol.

An element of the matrix is denoted by the corresponding small italic letter which has one or two subscripts. The first subscript indicates the row, the second subscript the column in which the element appears:

$$\underset{rs}{\mathrm{A}} = \mathrm{A} = \begin{Bmatrix} a_{11} & a_{12} & \cdots & a_{1s} \\ a_{21} & a_{22} & \cdots & a_{2s} \\ a_{31} & a_{32} & \cdots & a_{3s} \\ \cdots & \cdots & \cdots & \\ a_{r1} & a_{r2} & \cdots & a_{rs} \end{Bmatrix}$$

The *transpose* of matrix A contains the same elements as the original matrix but the columns and rows have been interchanged:

$$\underset{sr}{\mathrm{A}^T} = \mathrm{A}^T = \begin{Bmatrix} a_{11} & a_{21} & a_{31} & \cdots & a_{r1} \\ a_{12} & a_{22} & a_{32} & \cdots & a_{r2} \\ \cdots & \cdots & \cdots & \cdots & \\ a_{1s} & a_{2s} & a_{3s} & \cdots & a_{rs} \end{Bmatrix}$$

As a rule, we shall choose that arrangement as the original matrix in which there are more rows than columns.

Special matrices

Vector is a matrix which has only one column or, in the transposed arrangement, one row. The elements of a vector are called also *components*. We shall denote vectors by small Roman letters:

$$\underset{r1}{l} = l = \begin{Bmatrix} \ell_1 \\ \ell_2 \\ \ell_3 \\ .. \\ \ell_r \end{Bmatrix} \quad \text{or} \quad \underset{1r}{l^T} = l^T = \{\ell_1 \ell_2 \ell_3 \; ... \; \ell_r\}$$

Square matrix has as many columns as rows. A square matrix is *symmetric* if the elements are equal which are situated symmetrically with respect to the *main diagonal* (from upper left hand corner to lower right hand corner). The symmetric matrix is identical with it's own transpose:

$$N^T \equiv N \quad \text{if} \quad n_{ij} = n_{ji}$$

Upper triangular matrix is a square matrix in which the elements below the main diagonal are all $= 0$. Usually we shall denote the general triangular matrix by R, then $r_{ij} = 0$ when $i > j$. In practical computations, the symmetrical matrix is also written in form of a triangular matrix, in order to save time. In ambiguous cases, the main diagonal elements of a symmetrical matrix can be underlined.

Diagonal matrix is a symmetrical matrix in which all elements outside the main diagonal are $= 0$. In practical computations, however, the diagonal matrix is often written in form of a column, in order to save space.

Identity matrix is a diagonal matrix all elements of which are $= 1$. We shall use always the symbol I for the identity matrix.

Zero matrix 0 and *zero vector* o appear only in computation formulas; they are usually not written down because all elements are $= 0$.

An extensive matrix with many rows and columns can be divided by horizontal and vertical lines into *partitioned matrices*. Inversely, smaller matrices can be united to a *composed matrix*. In matrix operations, the partitioned matrices can be dealt with as elements of the original matrix or the original matrices as elements of the composed matrix.

12.2 *Matrix operations*

Addition and *subtraction* of matrices are possible when all matrices have the same dimension numbers; the sum or the difference are also matrices with the same dimension numbers:

$$\underset{rs}{A} + \underset{rs}{B} = \underset{rs}{C} \text{ when } a_{ij} + b_{ij} = c_{ij} \qquad (12.01)$$

A matrix is *multiplied* by a number when each element is multiplied by this number.

The *product of two matrices* is defined with the following restriction. The first factor (multiplicator) must have the same number of columns as the second factor (multiplicand) has rows; the product has as many rows as the first factor and as many columns as the second factor:

$$\underset{rt}{A}\ \underset{ts}{B} = \underset{rs}{X} \qquad (12.02a)$$

Each element of the product matrix is a sum of products:

$$x_{ij} = a_{i1}b_{1j} + a_{i2}b_{2j} + \ldots + a_{it}b_{tj} \qquad (12.02b)$$

The practical computation by the aid of a desk calculator can be carried out according to the following scheme.

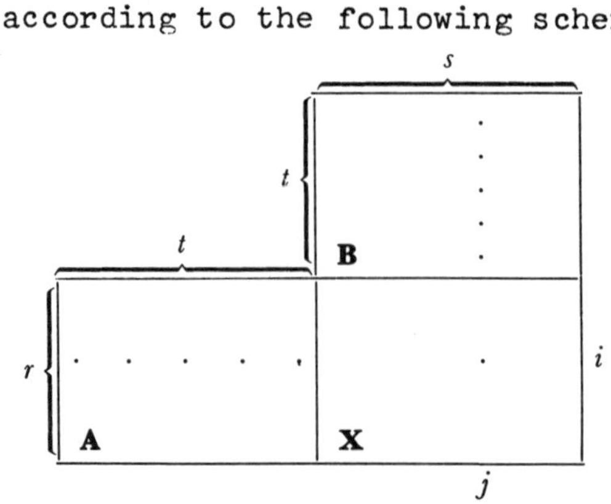

When the element x_{ij} must be computed, we can use two coins moving the first one step by step along the row i of the first matrix and, similarly, the second coin along the column j of the second matrix. Often the matrix is written down in the transposed form, then only the rows and columns must be interchanged. At each step of the movement, an accumulative multiplication between the elements shown by the coins is performed.

By the product of two matrices a third matrix can be multiplied. It can be shown easily that the matrix multiplication is *associative:*

$$ABC = (AB)C = A(BC) \tag{12.03}$$

It is also *distributive:*

$$A(B+C) = AB + AC \tag{12.04}$$

but generally it is *not commutative.* E.g., if the product AB can be computed as in formula (12.02a), the product BA has no meaning if $r \neq s$, and if $r = s$, the product BA usually is not equal to AB.

As an exercise, the reader should demonstrate and memorize the following special cases:

1. Product of a matrix and a column is a column;
2. Product of a row and a matrix is a row;
3. Product of a column and a row is a matrix;
4. Product of a row and a column is a number;
5. Product of a row vector and it's own transposed vector is a square sum;
6. When a matrix is multiplied by it's own transpose, the product is a symmetrical matrix with positive diagonal elements;
7. A product and all it's factor matrices can be transposed simultaneously, if the order of the factors is reversed at the same time:

   If $ABC = Z$ then $C^T B^T A^T = Z^T$
8. Product of a diagonal matrix and a rectangular matrix is a new matrix in which each row of the rectangular matrix is multiplied by a constant;
9. Product of a rectangular matrix and a diagonal matrix is a new matrix in which each column of the first matrix is multiplied

by a constant;

10. The identity matrix has no effect in a matrix multiplication:

$$IA = AI = A$$

Matrix division means the problem: a matrix product and one of the factors are given, find the other factor. We shall discuss first the case in which the unknown matrix is the second factor and the product is a column:

$$\underset{rs}{N} \underset{sl}{x} = \underset{rl}{u} \qquad (12.05)$$

According to the definitation of matrix multiplication, the elements *n* give *r* linear equations between *s* unknown elements. If *r* > *s*, then it usually impossible to find such a group of solutions which could satisfy all the equations. If *s* > *r*, there is an infinite number of possible solutions. Accordingly, a solution can be found only when *r* = *s* or when N is a square matrix. In addition, the determinant of the coefficients in such a set of equations must be ≠ 0. In fact, each square matrix forms also a determinant the numerical value of which can be computed by formula (12.17); if this value is = 0, the matrix is called *singular*. Matrix division is possible only when the given factor, called also the *divisor*, is a non-singular matrix.

If the known product matrix has several columns, the columns of the unknown factor can be computed one by one as in case (12.05). In case NX = U we have then a "division from the left", in case XN = U we have a "division from the right", but the latter case can be reversed to the former case by transpositions: $N^TX^T = U^T$. Anyway, we can not use the symbols of a customary division like X = U:N or X = U/N which do not allow the distinction between these cases. In this book, a special symbol is suggested:

Case NX = U, division from the left, $X = \overline{N}U$ (12.06)

Case XN = U, division from the right, $X = U\overline{N}$ (12.07)

The divisor is customarily "below the division line" but to the left or to the right, respectively, from the dividend.

In addition, the symbol $\overline{N}$ has another interpretation: it is the *inverse matrix* of N. If we have two division problems:

$$NX = I \quad \text{and} \quad YN = I \qquad (12.08)$$

and if we compute the product YNX in two different ways:

$$(YN)X = IX = X = Y(NX) = YI = Y$$

we can conclude that the problems (12.08) have only one solution if any. This solution is usually denoted by "N in power -1" or

$$X = Y = N^{-1} \qquad (12.09)$$

and called the inverse of N but, in the long formulas of adjustments, the symbol suggested in (12.06) and (12.07) offers certain advantages. As a definition, we can write also

$$N\bar{N} = \bar{N}N = I \qquad (12.10)$$

As an exercise, the reader should assure oneself of the following rules.

11. Inverse of a triangular matrix is also a triangular matrix.
12. Inverse of a diagonal matrix is diagonal; the corresponding elements of both matrices are inverse numbers.
13. If A, B, C... are not singular and ABC... = Z, then

$$\bar{Z} = \ldots\bar{C}\bar{B}\bar{A}$$

14. If $N x = o$, then either $x = o$ or N is singular.
15. If $B^T l = B^T w$ and B is not a square matrix, we can not conclude that $l = w$.

12.3 *Gauss' algorithm*

The practical procedures of matrix division are usually based on Gauss' algorithm which was explained in Chapter 8.1. In matrix notations, the non-singular square matrix which will be used as the divisor must first be expressed as a product of two triangular matrices the first of which is transposed and has the identity matrix as the main diagonal:

$$N = S^T R \qquad (12.11)$$

The division $N x = u$ can then be carried out in two steps:

$$S^T c = u \quad \text{or} \quad c = \bar{S}^T u \qquad (12.12)$$

$$R x = c \quad \text{or} \quad x = \bar{R} c \qquad (12.13)$$

The elements of the new triangular matrices can be computed one by one in an obvious order, which can be easily devised if we think that the matrix multiplication (12.11) is performed by the aid of the scheme given below:

Scheme
for matrix multiplication
$S^TR = N$

$$
\begin{array}{c|c}
 & \begin{array}{cccccccc}
r_{11} & r_{12} & \cdot & r_{1i} & \cdot & r_{1j} & \cdot & r_{1z} \\
 & r_{22} & \cdot & r_{2i} & \cdot & r_{2j} & \cdot & r_{2z} \\
 & & \cdot & \cdot & \cdot & \cdot & \cdot & \cdot \\
 & & & r_{ii} & \cdot & r_{ij} & \cdot & r_{iz} \\
 & & & & \cdot & \cdot & \cdot & \cdot \\
 & & & & & r_{jj} & \cdot & r_{jz} \\
 & & & & & & \cdot & \cdot \\
 & & & & & & & r_{zz}
\end{array} \\
\hline
\begin{array}{cccccccc}
1 & & & & & & & \\
s_{12} & 1 & & & & & & \\
\cdot & \cdot & \cdot & & & & & \\
s_{1i} & s_{2i} & \cdot & 1 & & & & \\
\cdot & \cdot & & \cdot & & & & \\
s_{1j} & s_{2j} & \cdot & s_{ij} & \cdot & 1 & & \\
\cdot & \cdot & \cdot & \cdot & \cdot & \cdot & \cdot & \\
s_{1z} & s_{2z} & \cdot & s_{iz} & \cdot & s_{jz} & \cdot & 1
\end{array} &
\begin{array}{cccccccc}
n_{11} & n_{12} & \cdot & n_{1i} & \cdot & n_{1j} & \cdot & n_{1z} \\
n_{21} & n_{22} & \cdot & n_{2i} & \cdot & n_{2j} & \cdot & n_{2z} \\
\cdot & \cdot & \cdot & \cdot & \cdot & \cdot & \cdot & \cdot \\
n_{i1} & n_{i2} & \cdot & n_{ii} & \cdot & n_{ij} & \cdot & n_{iz} \\
\cdot & \cdot & \cdot & \cdot & \cdot & \cdot & \cdot & \cdot \\
n_{j1} & n_{j2} & \cdot & n_{ji} & \cdot & n_{jj} & \cdot & n_{jz} \\
\cdot & \cdot & \cdot & \cdot & \cdot & \cdot & \cdot & \cdot \\
n_{z1} & n_{z2} & \cdot & n_{zi} & \cdot & n_{zj} & \cdot & n_{zz}
\end{array}
\end{array}
$$

We see that the first row of R is same as the first row of N and that the first column of S^T is the first column of N divided by $r_{11} = n_{11}$. After that we can compute in turn one row to R and one column to S^T. The general equations for the solution are

$$n_{ij} = s_{1i}r_{1j} + s_{2i}r_{2j} + \dots + r_{ij} \qquad \text{when } j \geq i \qquad (12.14)$$

$$n_{ji} = s_{1j}r_{1i} + s_{2j}r_{2i} + \dots + s_{ij}r_{ii} \qquad \text{when } j < i \qquad (12.15)$$

In (12.14), the only unknown is r_{ij}, in (12.15), s_{ij}. In the latter case, the last operation consists of a division by r_{ii} which, therefore, should never be $= 0$.

The complete theory of matrices which would take too much space here has shown that the determinant of N is = the product of all diagonal elements of R:

$$\det(N) = |N| = r_{11}r_{22}r_{33} \dots r_{zz} \qquad (12.16)$$

N is singular if, at least, one of these elements is zero; we have already stated that the division by N in such a case is impossible.

The division (12.12) by S^T can be performed at the same time and by the same procedure as the computation of R. The general equation reads

$$u_i = s_{1i}c_1 + s_{2i}c_2 + \dots + c_i \qquad (12.17)$$

During the solution of equations (12.14) and (12.17), we must transfer to the left hand side all other terms of the right hand side except the last one. Therefore, it is advisable to change immediately the sign of each element of S in the computational schemes. In addition, it is not necessary to write down the diagonal elements which form the identity matrix. In other words, we do not write down the elements of S but those of a special *fillmatrix*

$$T = I - S^T \qquad (12.18)$$

which can be written into the empty corner below the main diagonal of R. Both matrices R and T can be separated from each other by a stair line as already explained in Chapter 8,5 or in the Table VII/9.5. By the aid of T, equation (12.11) can be replaced by a direct formula

$$R = N + TR \qquad (12.19)$$

and the equation (12.12) by

$$c = u + Tc \qquad (12.20)$$

When N is not symmetrical, we still need equations (12.15). This case, however, is not usual in adjustments but the solution of this problem has usually been programmed for any type of automatic computers. When only a desk computer is available, it is advisable to use similar complete schemes as given in Chapter 8.2 because the best order in which the unknown elements could be computed can not always be seen beforehand. We shall give numerical examples in this case in Chapter 17.3.

N is always symmetrical for the solution of normal equations. In this case, we can write

$$N = S^T HS \qquad (12.21)$$

where H is the main diagonal of R:

$$R = HS \qquad (12.22)$$

In other words, $h_{ii} = r_{ii}$, $h_{ij} = 0$ when $i \neq j$. In this case, the long equation (12.15) can be replaced by (12.22) and (12.18) or by

$$t_{ji} = -\frac{r_{ij}}{r_{ii}} \tag{12.23}$$

The last step of the division (12.11) consists of the *back solution*. In an asymmetrical case, we can use the equation (12.13) which gives the elements x one by one but in the backward order. In a symmetrical case, we can write a new equation

$$Sx = d \tag{12.24}$$

for which we compute first the elements

$$d_i = \frac{c_i}{r_{ii}} \tag{12.25}$$

Then the back solution can be performed by formula

$$x^T = d^T + x^T T \tag{12.26}$$

The solution can be checked in the following way. When the row d^T is finished, we compute the product sum $d^T c$. When the row x^T is finished, we compute the product sum $x^T u$. By the aid of formulas (12.24) and (12.12) we can now prove that

$$d^T c = x^T S^T \bar{S}^T u = x^T u \tag{12.27}$$

Of course, a complete check is obtained by the multiplication according to formula (12.05).

12.4 *Inversion of a matrix*

When N is a general asymmetrical square matrix, the inverse matrix $\bar{N}$ can be found by a matrix division

$$NX = I \tag{12.28}$$

When the fillmatrix scheme is used, we write, after or instead of the column u, a diagonal of mere numbers 1. After or instead of the column c, the procedure indicated by (12.20) gives then

$$\bar{S}^T = I + T\bar{S}^T \tag{12.29}$$

Eventually, $X = \bar{N}$ is obtained by a back solution according to

$$RX = \bar{S}^T \tag{12.30}$$

When N is a symmetrical matrix, the formulas

$$\bar{\bar{R}} = \bar{\bar{S}}\bar{\bar{H}} \tag{12.31}$$

and

$$\bar{N} = \bar{\bar{R}}\bar{\bar{S}}^T \tag{12.32}$$

give the inverse matrix without a back solution.

In this case, $\bar{N}$ is symmetrical too. Therefore, as Banachiewicz has shown, it is not necessary to compute the elements of $\bar{\bar{S}}$. All we need are the diagonal elements of $\bar{\bar{R}}$, i.e. the inverse numbers $1/r_{ii}$. Then a back solution gives the elements of $\bar{N}$ according to

$$\bar{N} = \bar{\bar{R}} + \bar{N}T \tag{12.33}$$

$\bar{N}$ can be computed row by row starting from below; in this computation, only the diagonal and the zero elements of $\bar{\bar{R}}$ will be used. The elements in the upper righthand corner of $\bar{N}$ are obtained from the condition of symmetry. Numerical examples are given in Chapters 16.1 and 17.3.

In Chapter 8.6, we mentioned the *method of Cholesky* for the solution of normal equations; this solution is identical with a division by a symmetrical matrix N. We denote

$$K = \sqrt{H}S = \sqrt{\bar{H}}R \tag{12.34}$$

where $\sqrt{H}$ is a diagonal matrix the elements of which are square roots of the elements of the diagonal matrix H in (12.22). Then

$$N = K^T K \tag{12.35}$$

Denoting further

$$e = \sqrt{\bar{H}}c = \bar{\bar{K}}^T u \tag{12.36}$$

we obtain

$$Kx = e \tag{12.37}$$

which means the back solution according to the method of Cholesky. As a check, we obtain

$$e^T e = x^T u \tag{12.38}$$

The *method of Cholesky-Rubin* can be used for the division by a symmetrical matrix and for the inversion of the divisor matrix at the same time. After the column u, a diagonal of mere numbers 1 is written. Formula (12.36) gives the elements of e; after this column, the elements of $\bar{\bar{K}}^T$ are obtained by a similar procedure. Finally, we can compute without any back solutions

$$x^T = e^T\bar{K}^T \tag{12.39}$$

$$\bar{N} = \bar{K}\bar{K}^T \tag{12.40}$$

12.5 *Differential formulas*

Let the elements x_i $(i = 1, 2, \ldots. n)$ of a vector x mean free variables or parameters and the elements y_j $(j = 1, 2, \ldots m)$ of another vector y some mathematical functions of the parameters. Then we have $n \cdot m$ different partial derivatives

$$g_{ij} = \frac{\partial y_j}{\partial x_i} \tag{12.41}$$

Some definite numerical values of parameters form a known vector x_o and the corresponding values of the functions another known vector y_o. Further, we can compute the numerical values of (12.41) corresponding to x_o and y_o; these values form a rectangular matrix $\underset{nm}{G}$.

Let dx denote a vector which consists of small variations of the elements of x_o. Then the corresponding variations of the elements of y_o can be computed by a matrix multiplication

$$dy = G^T dx \tag{12.42}$$

Let now the elements z_h of a new vector z mean mathematical functions of the variables y_j. Then we have partial derivatives

$$f_{hj} = \frac{\partial z_h}{\partial y_j} \tag{12.43}$$

The numerical values of the elements z_h form a known vector z_o and those of the elements f_{hj} a rectangular matrix F. For the variations of the former, we obtain

$$dz = Fdy = FG^T dx \tag{12.44}$$

The matrix product FG^T represents, accordingly, the numerical values of the partial derivatives of z_h with respect to the parameters x_i.

Suppose now that $m = n$ in which case G is a square matrix. Suppose, moreover, that G is not singular. Then we obtain from (12.42)

$$dx = \bar{G}^T dy \tag{12.45}$$

The inverse matrix $\bar{G}$, accordingly, represents the partial derivatives of x_i with respect to y_j. As an example, cfr. formulas (3.31) and (3.36).

The *scalar product of two vectors* with an equal number of components is defined by formula

$$c = x^T y \tag{12.46}$$

If the elements of both vectors are variable, the variation of the product can be obtained by formula

$$dc = x^T dy + y^T dx \tag{12.47}$$

As a special case, we have the square sum of the components of one vector:

$$c = x^T x \tag{12.48}$$

with the variation

$$dc = 2x^T dx \tag{12.49}$$

In adjustments, we meet often a scalar product of the following form

$$c = x^T G x \tag{12.50}$$

where G is symmetric. The variation of c is then

$$dc = 2x^T G\, dx \tag{12.51}$$

12.6 *Matrix of weight coefficients*

Vector x means a multi-dimensional stochastic variate if the elements x_i are all random variables. We shall denote the variances of these components by σ_{ii} and the covariances of each pair of components by σ_{ij}. Because $\sigma_{ji} = \sigma_{ij}$, the variances and covariances can be written as elements of a symmetrical matrix, called *variance-covariance matrix* of x and denoted Σ_x.

Suppose that one stochastic variate has been chosen to serve as *weight unit*; the variance of this weight unit is here denoted by m^2. If all elements of Σ_x are divided by the constant m^2, we obtain a new symmetrical matrix which is called the matrix of *weight coefficients*. In this book, we shall denote the weight coefficients, according to the customary notation of adjustment but deviating from the general rule given in Chapter 12.1, by *capital* letters

$$Q_{ij} = \frac{\sigma_{ij}}{m^2} \tag{12.52}$$

The matrix itself is denoted by Q_x or $\bar{P}_x$. If this matrix is not singular it has an inverse matrix $\bar{Q}_x$ or P_x which can be called *weight matrix* of x.

If the components x_i are independent of each other, the covariances σ_{ij} are all $= 0$ and the matrices Q_x and P_x consist of mere diagonal matrices. The elements of P_x, in this case, are *weights* p_i of the components x_i, and those of Q_x are *inverse weights* $1/p_i$.

If the components x_i are correlated, the matrices Q_x and P_x can be complete symmetrical matrices. Also in this case, we call the diagonal elements of Q_x *inverse weights of* x_i and the inverse numbers of these elements *weights of* x_i but *the diagonal elements of* P_x*, in this case, are not identical with the weights.* The other elements of Q_x are called *cofactors.*

Returning to formula (3.10), we shall now transcribe it using the matrix notations. We see that the elements

$$\begin{matrix} \sigma_{xx} & \sigma_{xy} & \sigma_{xz} \\ \sigma_{xy} & \sigma_{yy} & \sigma_{yz} \\ \sigma_{xz} & \sigma_{yz} & \sigma_{zz} \end{matrix}$$

are in this formula already in the order of the matrix arrangement. The partial derivatives $\frac{\partial u}{\partial i}, \frac{\partial y}{\partial i}$ $(i = x, y, z\ldots)$ represent two columns of matrix G as defined by formula (12.41). We realize that by each element $\frac{\partial u}{\partial i}$ of one column has been multiplied all products of a sum

$$\sigma_{ij} \frac{\partial y}{\partial j} \quad (j = x, y, z\ldots)$$

which is one element of the matrix product $\Sigma_x G$. The entire sum on the right hand side of (3.10) is one element of the matrix product $G^T \Sigma_x G$. These computations can be illustrated by the following schem

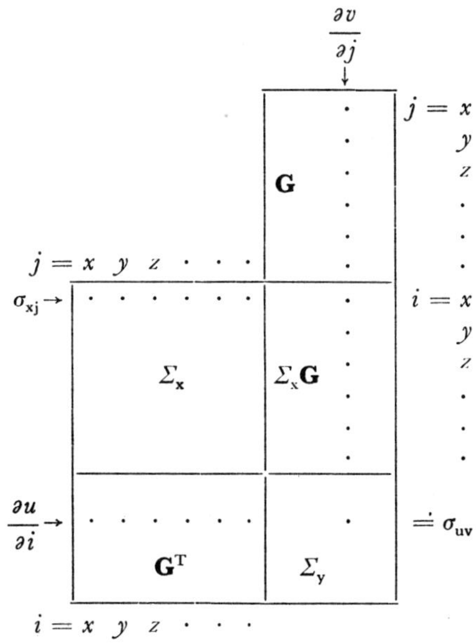

Denoting the functions $u, v, \ldots$ by vector $\mathbf{y}$ we can write the general law of the error propagation in matrix form

$$\Sigma_{\mathbf{y}} = G^T \Sigma_{\mathbf{x}} G \tag{12.53}$$

Dividing both sides of (12.53) by m^2 we obtain

$$Q_{\mathbf{y}} = G^T Q_{\mathbf{x}} \, G \tag{12.54}$$

Later, we shall have many applications of the important couple of formulas, (12.42) and (12.54).

We conclude this section with an interesting theorical discussion which hardly has any practical applications. Suppose that the components of $\mathbf{x}$ are all uncorrelated normal variables with unit weight. Then the variance-covariance matrix of x is = identity matrix I. Let now be

$$\mathbf{y} = G\,\mathbf{x} \tag{12.55}$$

where G is not singular. Then we have

$$\mathbf{x} = \bar{G}\,\mathbf{y} \tag{12.56}$$

The variance-covariance matrix of $\mathbf{y}$ is, according to (12.53),

$$\Sigma = GG^T \tag{12.57}$$

which, at the same time, is the matrix of weight coefficients. Then

$$P = \bar{G}^T\bar{G} \tag{12.58}$$

We want to study the distribution of **y** because it represents a general multi-dimensional *normal* variate with unequally weighted and correlated components.

Formula (12.55) represents the "projection" of an n - dimensional space **x** into another n - dimensional space **y**. Taking a small rectangular "volume" from the former:

$$dt_x = dx_1\ dx_2 \ldots dx_n \tag{12.59}$$

we have in the latter a non-rectangular volume

$$dt_y = |\mathbf{G}|\ dt_x \tag{12.60}$$

The probability that **x** is in the volume (12.59) is

$$dp = \phi(\mathbf{x})dt_x \tag{12.61}$$

where

$$\phi(\mathbf{x}) + (2\pi)^{-\frac{n}{2}} \exp(-\tfrac{1}{2}\ \mathbf{x}^T\mathbf{x}) \tag{12.62}$$

This formula is a simple generalization of formula (11.57). By the aid of (12.56), (12.58) and (12.60), we can now conclude that

$$\Psi(\mathbf{y}) = \frac{\exp(-\frac{1}{2}\mathbf{y}^T\mathbf{P}\mathbf{y})}{\sqrt{(2\pi)^n|\mathbf{P}|}} \tag{12.63}$$

gives the *probability density function for a normal distribution in a n-dimensional space.* Formula (11.58) obviously is the special case $n = 2$ of this general formula.

In Chapter 14.2 we shall discuss an inverted problem: when correlated variables (here **y**) are given, we shall find an orthogonal transformation (12.56) which makes the new variables **x** uncorrelated.

13 *THE USE OF MATRICES IN ADJUSTMENTS*

13.1 *Method A: adjustment by parameters*

The unknown parameters are denoted by vector **x** with *u* components using subscripts for the different sets of numerical values:

$\mathbf{x}_o$ the initial approximate values chosen arbitrarily before the adjustment,

$\mathbf{x}_a$ the definitive values after the adjustment.

The differences of these two sets, or the variations of the parameters, are denoted without subscripts:

$$\mathbf{x} = \mathbf{x}_a - \mathbf{x}_o \qquad (13.01)$$

In some special methods, we shall use consecutive numbers as indices:

$\mathbf{x}_i$ (i = 1,2,3 ...) the successive variations for an iteration or the groups of elements (partitioned vectors).

The measured quantities are denoted by vector **l** with *n* components using following subscripts:

$\mathbf{l}_b$ the measured values,

$\mathbf{l}_o$ the values computed from the values $\mathbf{x}_o$ of parameters,

$\mathbf{l}_a$ the adjusted values.

The differences of these sets are denoted without subscripts:

$$\mathbf{l} = \mathbf{l}_o - \mathbf{l}_b \qquad (13.02)$$

$$\mathbf{v} = \mathbf{l}_a - \mathbf{l}_b \qquad (13,03)$$

In special methods, consecutive numbers can be used as subscripts.

The partial derivatives of the measured quantities with respect to the parameters are denoted by matrix **A** with *n* rows and *u* columns. The elements of **A**

$$a_{ij} = \frac{\partial \ell_{o,i}}{\partial x_{o,j}} \qquad (13.04)$$

are supposed to be constants, corresponding to the initial approximate values of the variables. Then we can write a matrix equation

$$l_a = l_o + A\,(x_a - x_o)$$

or, using notations (13.01) - (13.03):

$$v = A\,x + l \tag{13.05}$$

Formula (13.05) represents the *observation equations* in matrix form.

Usually we suppose that the observations are uncorrelated. Then we must estimate the weight p_{ii} of each observation $\ell_{b,i}$ and consider these weights as elements of a diagonal matrix **P**. The *principle of least squares*, $[pvv]$ = min., can then be expressed in matrix form:

$$v^T P\,v = \text{minimum} \tag{13.06}$$

If the observations have equal weights, we can put **P** = **I** and write $v^T v$ = min.

If the observations are correlated in such a way that both the variances and the covariances of the observed values can be estimated - or, at least, the relative values of them - we write first the matrix of weight coefficients **Q** and replace the weight matrix by $P = \bar{Q}$. This theoretical case, however, can not be applied in practice except in some special methods; cfr.Chapter 15.5.

Inserting (13.05) into (13.06) we obtain

$$v^T P\,v = x^T A^T P\,A\,x + 2\,l^T P\,A\,x + l^T P\,l$$

Introducing abbreviating notations

$$N = A^T P\,A \tag{13.07}$$

$$u = A^T P\,l \tag{13.08}$$

we can write

$$v^T P\,v = x^T N\,x + 2\,x^T u + l^T P\,l \tag{13.09}$$

The minimum value of this function is found when the partial derivatives of it with respect to all elements of **x** are made to vanish. According to Chapter 12.5, this condition gives the *normal equations*

$$N\,x + u = o \tag{13.10}$$

The solution of the normal equations can be written in matrix form

$$\mathbf{x} = -\bar{\mathbf{N}}\,\mathbf{u} \qquad (13.11)$$

For the numerical solution, however, it is not necessary to compute the elements of $\bar{\mathbf{N}}$ but a matrix division according to Chapter 12.3 is sufficient. We repeat the necessary computation formulas:

$$\mathbf{R} = \mathbf{N} + \mathbf{T}\,\mathbf{R} \qquad (13.12)$$

$$\mathbf{c} = \mathbf{u} + \mathbf{T}\,\mathbf{c} \qquad (13.13)$$

$$t_{ji} = -\frac{r_{ij}}{r_{ii}} \qquad (13.14)$$

$$-d_i = -\frac{c_i}{r_{ii}} \qquad (13.15)$$

$$\mathbf{x}^T = -\mathbf{d}^T + \mathbf{x}^T\,\mathbf{T} \qquad (13.16)$$

Inserting (13.10) into (13.09) we obtain

$$\mathbf{v}^T\mathbf{p}\,\mathbf{v} = \mathbf{l}^T\mathbf{P}\,\mathbf{l} + \mathbf{x}^T\mathbf{u} \qquad (13.17)$$

which can be used as a control of computations. Furthermore, using the triangular matrix $S = I - T^T$, we can write

$$\mathbf{N} = \mathbf{S}^T\,\mathbf{R} \qquad (13.18)$$

$$\mathbf{u} = \mathbf{S}^T\,\mathbf{c} \qquad (13.19)$$

$$\mathbf{x} = -\bar{\mathbf{S}}\,\mathbf{d} \qquad (13.20)$$

$$\mathbf{x}^T\mathbf{u} = -\mathbf{d}^T\mathbf{c} \qquad (13.21)$$

The right hand side of (13.21) can be computed before the "back solution" by formula (13.16), the left hand side after it, respectively. The elements of $\mathbf{x}$ give then the residuals $\mathbf{v}$ according to (13.05) and, eventually, the left hand side of (13.17) can be evaluated.

If equation (13.05) is multiplied by $\mathbf{A}^T\mathbf{P}$, the right hand side becomes identical with the left hand side of (13.10). Accordingly,

$$\mathbf{A}^T\mathbf{P}\,\mathbf{v} = \mathbf{o} \qquad (13.22)$$

which can be used for additional controls.

13.2 *Method B: adjustment by correlates*

Formula (13.05) is the matrix representation of n linear equations with u unknown parameters. If these parameters are eliminated from the equations, there remain $r = n - u$ equations which contain the unknown residuals only. In matrix form, the elimination can be represented by a multiplication. Equation (13.05) is multiplied by a matrix B^T such that

$$\mathbf{B}^T\mathbf{A} = \mathbf{0} \tag{13.23}$$

Denoting

$$\mathbf{B}^T\mathbf{l} = -\mathbf{w} \tag{13.24}$$

we obtain

$$\mathbf{B}^T\mathbf{v} + \mathbf{w} = \mathbf{o} \tag{13.25}$$

which is the matrix form for the condition equations.

In practice, this operation is usually an unnecessary detour. In fact, the method of condition equations has a practical significance only when these equations can be formed immediately without equations (13.05). For this purpose, we must find r functions of observed quantities. These functions must be independent of each other and their numerical values must be theoretically known even if the parameters are unknown. We denote

$\mathbf{w}_a$ the theoretical values of these functions

$\mathbf{w}_b$ the values computed from the observations

The differences

$$\mathbf{w} = \mathbf{w}_b - \mathbf{w}_a \tag{13.26}$$

are called *misclosures*.

The partial derivatives of these functions with respect to the observed quantities are denoted by matrix B with n rows and r columns. It's elements

$$b_{ij} = \frac{\partial w_{b,j}}{\partial \ell_{b,i}} \tag{13.27}$$

are supposed to be constants, corresponding to the observed values of the variables. In this way, we obtain immediately

$$\mathbf{w}_a = \mathbf{w}_b + \mathbf{B}^T(\mathbf{l}_a - \mathbf{l}_b)$$

or, using notations (13.03) and (13.26), equations (13.25).

These equations are satisfied by an infinite number of different vectors **v** but we shall find such a vector which, at the same time, satisfies the condition (13.05). For this purpose, we construct a function

$$\frac{1}{2}\,\mathbf{v}^T\mathbf{P}\,\mathbf{v} - (\mathbf{w}^T + \mathbf{v}^T\mathbf{B})\,\mathbf{k} = \text{minimum}$$

in which **k** is an unknown constant vector. The minimum value of this function is found when the partial derivatives of it with respect to all elements of **v** vanish. According to Chapter 12.5, this condition gives

$$\mathbf{P}\,\mathbf{v} = \mathbf{B}\,\mathbf{k} \tag{13.28}$$

or

$$\mathbf{v} = \bar{\bar{\mathbf{P}}}\,\mathbf{B}\,\mathbf{k} \tag{13.29}$$

Inserting (13.29) into (13.25) and introducing an abbreviating notation

$$\mathbf{M} = \mathbf{B}^T\,\bar{\bar{\mathbf{P}}}\,\mathbf{B} \tag{13.30}$$

we obtain the *normal equations*

$$\mathbf{M}\,\mathbf{k} + \mathbf{w} = \mathbf{o} \tag{13.31}$$

The solution of (13.31) can be written in matrix form

$$\mathbf{k} = -\,\bar{\bar{\mathbf{M}}}\,\mathbf{w} \tag{13.32}$$

For the numerical computations, however, we use formulas

$$\mathbf{R} = \mathbf{M} + \mathbf{T}\,\mathbf{R} \tag{13.33}$$

$$\mathbf{c} = \mathbf{w} + \mathbf{T}\,\mathbf{c} \tag{13.34}$$

$$t_{ji} = -\frac{r_{ij}}{r_{ii}} \tag{13.35}$$

$$-d_i = -\frac{c_i}{r_{ii}} \tag{13.36}$$

$$\mathbf{k}^T = -\,\mathbf{d}^T + \mathbf{k}^T\,\mathbf{T} \tag{13.37}$$

When vector **k** is found, formulas (13.28) and (13.29) give the residuals **v**.

Inserting (13.29) into the matrix product $\mathbf{v}^T\mathbf{P}\,\mathbf{v}$, we obtain

$$\mathbf{k}^T\mathbf{B}^T\bar{\bar{\mathbf{P}}}\,\mathbf{P}\,\bar{\bar{\mathbf{P}}}\,\mathbf{B}\,\mathbf{k} = \mathbf{k}^T\mathbf{B}^T\bar{\bar{\mathbf{P}}}\,\mathbf{B}\,\mathbf{k} = \mathbf{k}^T\mathbf{M}\,\mathbf{k}$$

or, according to (13.31)

$$\mathbf{v}^T\mathbf{P}\,\mathbf{v} = -\,\mathbf{k}^T\mathbf{w} \tag{13.38}$$

Using the triangular matrix $S = I - T^T$, we can write

$$M = S^T R \tag{13.39}$$

$$\mathbf{w} = S^T\ \mathbf{c} \tag{13.40}$$

$$\mathbf{k} = -\ \bar{S}\ \mathrm{d} \tag{13.41}$$

$$\mathbf{v}^T P\ \mathbf{v} = \mathrm{d}^T\ \mathbf{c} \tag{13.42}$$

The last formula (13.42) gives a double check for the residuals **v**. In the *method of Cholesky*, see Chapter 12.4, we have

$$M = K^T K \tag{13.43a}$$

$$\mathbf{w} = K^T \mathbf{e} \tag{13.43b}$$

$$\mathbf{k} = -\ \bar{K}\ \mathbf{e} \tag{13.43c}$$

$$\mathbf{v}^T P\ \mathbf{v} = \mathbf{e}^T \mathbf{e} \tag{13.44}$$

13.3 *Weight coefficients of the parameters*

In the following chapters, we shall use repeatedly formula (12.54), which in the case of linear functions can be written in form

$$\text{if} \quad \mathbf{y} = \mathbf{y}_o + G^T \mathbf{x} \tag{13.45}$$

$$\text{then} \quad Q_y = G^T Q_x G$$

We start from vector $\mathbf{l}_b$ the weight coefficients of which are denoted by symbol $\bar{P}$ without any index. In practice, $\bar{P}$ usually is a diagonal matrix, in theory, it can be any non-singular symmetrical matrix.

First, we take vector **u** which represents the constant terms of the normal equations in **A** - method:

$$\mathbf{u} = A^T P\ \mathbf{l} \tag{13.08}$$

According to (13.45), we obtain

$$Q_u = A^T P\ \bar{P}\ P\ A = A^T P\ A = N \tag{13.46}$$

Taking now vector **x** which represents the parameters:

$$\mathbf{x} = -\ \bar{N}\ \mathbf{u} \tag{13.11}$$

we obtain

$$Q_x = \bar{N}\, N\, \bar{N} = \bar{N} \qquad (13.47)$$

Accordingly, *the weight coefficients of the parameters can be found by inverting the matrix of the coefficients of the normal equations.* Especially, the inverse weight of the last parameter x_u is $1/r_{uu}$, hence, *the weight of* x_u is

$$p(x_u) = r_{uu} = h_{uu} \qquad (13.48)$$

Often the standard errors and, some times, also the weight coefficients are needed for a few parameters only. If these parameters are put on the last places, the weight of the last one is found in the course of the solution of normal equations without any additional computations and the matrix of the necessary weight coefficients by interrupting the back solution procedure of matrix inversion at the proper step.

13.4 *Standard error of weight unit*

For the vector of misclosures we had formula

$$w = -B^T l \qquad (13.24)$$

According to (13.45), we obtain

$$Q_w = B^T \bar{P}\, B = M \qquad (13.49)$$

Taking now the correlates in form

$$k = -\bar{M}\, w \qquad (13.32)$$

we obtain the weight coefficients of the correlates

$$Q_k = \bar{M} \qquad (13.50)$$

From (13.40) we obtain

$$c = \bar{S}^T w \qquad (13.51)$$

and from (13.36)

$$d = \bar{H}\, c \qquad (13.52)$$

For the weight coefficients we obtain, respectively,

$$Q_c = \bar{S}^T M\, \bar{S} = \bar{S}^T S^T H\, S\, \bar{S} = H \qquad (13.53)$$

$$Q_d = \bar{H}\, H\, \bar{H} = \bar{H} \qquad (13.54)$$

The formulas given above have no practical significance. However, if the normal equations of the B - method are solved by the method of Cholesky, the constant terms of the reduced normal equations consist of vector **e**, see (12.36):

$$\mathbf{e} = \sqrt{\bar{\mathrm{H}}}\ \mathbf{c} \tag{13.55}$$

The matrix of the weight coefficients is, accordingly

$$\mathbf{Q}_{\mathbf{e}} = \sqrt{\bar{\mathrm{H}}}\ \mathrm{H}\ \sqrt{\bar{\mathrm{H}}} = \mathrm{I} \tag{12.56}$$

This formula shows that the components of vector **e** can be regarded as uncorrelated stochastic variates which all have the weight 1. At the same time, they form the greatest possible sample of such variates which can be obtained from the measurements on the basis of the accepted theory. The expected values of these components are = 0, if this basic theory is reliable; in this respect the components of **e** correspond to the "true errors" in formula (4.07). We conclude, that this formula then gives the best unbiased estimate for all observations which have weight 1:

$$m^2 = \frac{[ee]}{r} \tag{13.57}$$

According to (13.44), $[ee] = \mathbf{e}^T\mathbf{e} = \mathbf{v}^T\mathbf{P}\ \mathbf{v} = [pvv]$. In this way, we have found an elegant proof for the formula

$$m^2 = \frac{[pvv]}{r} = \frac{[pvv]}{n - u} \tag{13.58}$$

which we have already mentioned in various connections. This proof was given by *Tienstra*; the textbooks usually give a longer proof which is based on the spur of matrix J, cfr. the following chapter.

13.5 *Weight coefficients of the adjusted observations*

The adjusted values of the observed quantities are represented by vector $\mathbf{l}_a = \mathbf{l}_b + \mathbf{v} = \mathbf{l}_o - \mathbf{l} + \mathbf{v}$. Because $\mathbf{l}_o$ is an arbitrary constant vector, the weight coefficients of $\mathbf{l}_a$ are the same as those of

$$\mathbf{v} - \mathbf{l} = \mathbf{A}\ \mathbf{x} \tag{13.59}$$

According to (13.47), we obtain

$$Q_{1-v} = \mathbf{A}\,\bar{\mathbf{N}}\,\mathbf{A}^T \tag{13.60}$$

In method *B*, we had $\mathbf{v} = \bar{\mathbf{P}}\,B\,\mathbf{k}$ and, according to (13.50),

$$Q_v = \bar{\mathbf{P}}\,B\,\bar{\mathbf{M}}\,B^T\bar{\mathbf{P}} \tag{13.61}$$

Introducing new symbols

$$J_A = \mathbf{A}\,\bar{\mathbf{N}}\,\mathbf{A}^T\mathbf{P} \tag{13.62}$$

$$J_B = \mathbf{P}\,B\,\bar{\mathbf{M}}\,B^T \tag{13.63}$$

we can write

$$Q_{1-v} = J_A\bar{\mathbf{P}} \tag{13.64}$$

$$Q_v = J_B\bar{\mathbf{P}} \tag{13.65}$$

$$Q_{1-v} = \bar{\mathbf{P}} - J_B\,\bar{\mathbf{P}} \tag{13.66}$$

By the aid of these symbols, we obtain from (13.59), (13.11), and (13.08)

$$\mathbf{l} - \mathbf{v} = J_A\mathbf{l} \tag{13.67}$$

and from (13.29), (13.32), and (13.24)

$$\mathbf{v} = J_B\mathbf{l} \tag{13.68}$$

Because both methods of adjustment must give same results, we conclude that

$$J_A + J_B = I \tag{13.69}$$

J_A and J_B are square matrices with n rows and n columns. If $\mathbf{P} = \mathbf{I}$, they are symmetrical matrices. Anyway, they are singular matrices which cannot be inverted. They are known by the name *idempotent matrices* because, in some special cases, they have no effect in matrix products. E.g., we can see that

$$J_A\mathbf{A} = \mathbf{A} \tag{13.70}$$

$$B^TJ_B = B^T \tag{13.71}$$

Furthermore, $J_AJ_A = J_A$, $J_BJ_B = J_B$ etc.

In this connection, we may mention that the term *spur* of a square matrix means the sum of it's diagonal elements. It could be shown that the spur of J_A is u (the number of parameters) and the spur of J_B is r (the number of conditions). From formula (13.64) we can conclude that an adjustment increases the weights of the observed quantities by factor $\frac{n}{u}$, in average.

13.6 *Weight coefficients of functions*

In the course of an adjustment, we may need adjusted values and standard errors for quantities in addition to the parameters and the observed values. E.g., in a triangulation where the coordinates serve as parameters and the angles as observed quantities, we may need also the lengths of some sides.

We shall denote these additional quantities by vector $\mathbf{y}$ and express them as functions either of the parameters or as functions of the observed quantities. The numerical values of these functions are denoted by

- $\mathbf{y}_o$ when computed from the approximate values of parameters
- $\mathbf{y}_b$ when computed from the observations
- $\mathbf{y}_a$ when computed after the adjustment

The partial derivatives of these functions are denoted by matrices $\mathbf{G}$ and $\mathbf{F}$ with elements

$$g_{ij} = \frac{\partial y_{o,j}}{\partial x_{o,i}} \tag{13.72}$$

$$f_{ij} = \frac{\partial y_{b,j}}{\partial \ell_{b,i}} \tag{13.73}$$

As shown in Chapter 12.5, the partial derivatives of formula (13.04) give the functional dependence of $\mathbf{G}$ and $\mathbf{F}$:

$$\mathbf{G} = \mathbf{A}^T\mathbf{F} \tag{13.74}$$

We start from the differential formulas

$$\mathbf{y}_a = \mathbf{y}_o + \mathbf{G}^T\mathbf{x} \tag{13.75}$$

$$\mathbf{y}_a = \mathbf{y}_b + \mathbf{F}^T\mathbf{v} = \mathbf{y}_o - \mathbf{F}^T(\mathbf{l} - \mathbf{v}) \tag{13.76}$$

Because $\mathbf{y}_o$ is a constant vector, (13.75) gives

$$Q_{y_a} = G^T\bar{N}\,G \tag{13.77}$$

which gives immediately the weight coefficients of the adjusted values of functions $\mathbf{y}$ if $\bar{\mathbf{N}}$ has been readily computed. If the elements of $\bar{\mathbf{N}}$ are not necessary for other purposes, we can compute matrix (13.77) in an easier way. Denoting

$$\bar{N} = \bar{S}\,\bar{H}\,\bar{S}^T \tag{13.78}$$

$$\mathbf{C} = \bar{S}^T\mathbf{G} \tag{13.79}$$

$$D^T = G^T\bar{S}\,\bar{H} \tag{13.80}$$

we compute the elements of C and D in the course of the solution of normal equations:

$$\mathbf{C} = \mathbf{G} + \mathbf{T}\,\mathbf{C} \tag{13.81}$$

$$d_{ij} = \frac{c_{ij}}{r_{ii}} \tag{13.82}$$

In the scheme of the numerical computations, matrix G is written on the prolonged rows of N and **u**; then C is obtained on the prolonged rows of R and **c** by similar operations as those, cfr. (13.12) and (13.13), and then D^T in prolonged columns of T and $\mathbf{d}^T$. Eventually, we compute

$$Q_{y_a} = D^T C \tag{13.83}$$

In method *B*, (13.76) gives $Q_{y_a} = \mathbf{F}^T Q_{l-v}\mathbf{F}$ which, according to (13.66), gives first

$$Q_{y_a} = \mathbf{F}^T\bar{P}\,\mathbf{F} - \mathbf{F}^T\bar{P}\,B\,\bar{M}\,B^T\bar{P}\,\mathbf{F} \tag{13.84}$$

Introducing a new symbol

$$E = B^T\bar{P}\,\mathbf{F} \tag{13.85}$$

(13.84) can be written in a shorter form

$$Q_{y_a} = \mathbf{F}^T\bar{P}\,\mathbf{F} - E^T\bar{M}\,E \tag{13.86}$$

Matrix E requires some additional computation at the same time and by similar operations as the computation of **M**. Then the rest of computations can be performed using formulas

$$\mathbf{C} = \mathbf{E} + \mathbf{T}\,\mathbf{C} \qquad (13.87)$$

$$d_{ij} = \frac{c_{ij}}{r_{ii}} \qquad (13.88)$$

$$\mathbf{Q}_{\mathbf{y}_a} = \mathbf{F}^T\bar{\mathbf{P}}\,\mathbf{F} - \mathbf{D}^T\mathbf{C} \qquad (13.89)$$

If parameters $\mathbf{x}$ and their weight coefficients are required in an adjustment which is performed by method B, they must be expressed as functions of some selected observations and used as functions $\mathbf{y}$ of this chapter.

Numerical examples of the computations explained above will be given in Chapters 16.2 and 16.3.

13.7 *Alphabetical list of matrix symbols*

The following list contains all symbols which in this book have a specified meaning. The figures in parentheses refer to the formulas which give the definition or the best method of computation for the matrix in question.

	Subscripts		Dimension numbers
A	Adjustment by parameters	f	Number of functions
B	Adjustment by correlates	n	" " observations
a	Adjusted value	r	" " conditions
b	Observed value	r	" " rows
m	Mean value	s	" " columns
t	Theoretical value	u	" " parameters
o, 1, 2,...	Successive approximations		
1, 2, 3,...	Successive groups		
i, j, k,...	Indefinite subscripts		

Vectors

c_u, c_r	Constant terms of the reduced normal equations	(13.13),(13.34)
d_u, d_r	Constant terms of the fill matrix equations	(13.15),(13.36)
e_u, e_r	Constant terms of the Cholesky equations	(13.55)
k_r	Correlates	(13.37)
l_n	Differences $l_o - l_b$ (observed quantities)	(13.02)
o_r	Zero vector	Chapter 12.1
u_u	Constant terms of the normal equations in method A	(13.08)
v_n	Differences $l_a - l_b$ (residuals)	(13.03),(13.05),(13.29)
w_r	Misclosures	(13.24),(13.26)
x_u	Differences $x_a - x_o$ (parameters)	(13.01)
y_f	Functions of the parameters or of the observed quantities	Chapter 13.6
z_r	Intermediate quantities	Chapter 15.2

Matrices

A_{nu}	Coefficients of the observation equations	(13.04)
B_{nr}	Coefficients of the condition equations	(13.27)
C_{uf}, C_{rf}	Reduced coefficients of the functions	(13.81)
D_{uf}, D_{rf}	Fill matrix coefficients of the functions	(13.82)
E_d	Coefficients of the functions of correlates	(13.85)
F_{nf}	" " " " " observed quantities	(13.73)
G_{uf}	" " " " " parameters	(13.72)
H_{uu},H_{rr}	Diagonal elements of R	(12.22)

$\mathbf{I}_{ss}$	Identity matrix	Chapter 12.1
$\mathbf{J}_{nn}$	Idempotent matrix	(13.62),(13.63)
$\mathbf{K}_{uu}$, $\mathbf{K}_{rr}$	Coefficients of the Cholesky equations	(12.35)
$\mathbf{M}_{rr}$	" " " normal equations in method *B*	(13.30)
$\mathbf{N}_{uu}$	" " " " " " " A	(13.07)
$\mathbf{O}_{rs}$	Zero matrix	Chapter 12.1
$\mathbf{P}_{ss}$	Weight matrix	Chapter 12.5
$\mathbf{Q}_{ss}$	Weight coefficients	(13.83),(13.89)
$\mathbf{R}_{uu}$, $\mathbf{R}_{rr}$	Coefficients of the reduced normal equations	(13.12),(13.33)
$\mathbf{S}_{uu}$, $\mathbf{S}_{rr}$	" " " end equations	(13.18),(13.39)
$\mathbf{T}_{uu}$, $\mathbf{T}_{rr}$	Fillmatrix $\mathbf{I} - \mathbf{S}^T$	(13.14),(13.35)
$\mathbf{Z}_{yz}$	Intermediate matrix (Zwischenkorrelaten of Boltz)	(15.11)
$\mathbf{\Sigma}_{ss}$	Covariance matrix	Chapter 12.5
$\mathbf{A}^T$	Transpose of $\mathbf{A}$	Chapter 12.1
$\bar{\mathbf{N}}$	Inverse of $\mathbf{N}$	(12.33)
$\lvert\mathbf{N}\rvert$	Determinant of $\mathbf{N}$	(12.16)

14 *THE ORTHOGONAL TRANSFORMATIONS*

14.1 *Error ellipse*

In some problems, we may have two special parameters x and y for which an adjustment has given, in addition to the adjusted values, also the weight coefficients Q_{xx}, Q_{yy} and Q_{xy}. Besides, we may have an estimate m for the standard error of unit weight. The standard errors of x and y are then

$$m_x = m\sqrt{Q_{xx}}$$
$$m_y \quad m\sqrt{Q_{yy}}$$

In a graphical representation, the adjusted values x_a and y_a serve as coordinates of a point on xy-plane. The standard errors describe a rectangle

$$x_a - m_x < x < x_a + m_x$$

$$y_a - m_y < y < y_a + m_y$$

around this point; in this way we have a rough illustration about the two dimensional dispersion of the possible solutions for our problem. In a mathematical idealization, however, the rectangle is replaced by an ellipse to which the sides of the rectangle are tangents. In fact, there are several possible ellipses inside such a rectangle, each having different directions as main axes. We shall show here how the cofactor Q_{xy} defines the direction of one special ellipse which is called the *error ellipse* of the solution x_a, y_a.

First, we perform an *orthogonal transformation* of the parameters or, in our geometrical interpretation, a rotation of the rectangular system of coordinate axes in a multidimensional space. In this two dimensional case we have simply

$$u = x \cos t + y \sin t \qquad (14.01)$$
$$v = y \cos t - x \sin t$$

Here t means the direction angle of u-axis in the x, y-system.

Now we can define the standard error in direction t by

$$m_u = m \quad Q_{uu} \qquad (14.02)$$

where Q_{uu} can be computed according to (12.54);

$$Q_{uu} = Q_{xx}\cos^2 t + Q_{yy}\sin^2 t + 2Q_{xy}\sin t \cos t \qquad (14.03)$$

cfr. also (3.07). Using well known relations

$$\sin 2t = 2 \sin t \cos t$$
$$\cos 2t = \cos^2 t - \sin^2 t$$
$$1 = \cos^2 t + \sin^2 t$$

we obtain

$$Q_{uu} = \frac{1}{2}(Q_{xx} + Q_{yy}) + \frac{1}{2}(Q_{xx} - Q_{yy})\cos 2t + Q_{xy}\sin 2t \quad (14.04)$$

This quantity reaches it's extreme values when

$$\frac{dQ_{uu}}{dt} = 2\,Q_{xy}\cos 2t - (Q_{xx} - Q_{yy})\sin 2t = 0 \qquad (14.05)$$

The solution of this equations gives

$$\tan 2t = \frac{2\,Q_{xy}}{Q_{xx} - Q_{yy}} \qquad (14.06)$$

We can write also

$$\sin 2t = \frac{2\,Q_{xy}}{W} \qquad (14.07)$$

$$\cos 2t = \frac{Q_{xx} - Q_{yy}}{W}$$

where

$$W = \sqrt{(Q_{xx} - Q_{yy})^2 + 4\,Q_{xy}^{\;2}} \qquad (14.08)$$

Inserting these expressions into (14.04), we obtain

$$Q_{uu} = \frac{1}{2}(Q_{xx} + Q_{yy} + W) \qquad (14.09)$$

If W is taken with the positive sign, (14.09) gives a maximum value, the negative sign would give the minimum value. The corresponding values of t can be obtained without bias from (14.07); the two values of W gives two opposite directions for $2t$, or two rectangular directions for t. Obviously, the negative value W can be ascribed to v - axis:

$$Q_{vv} = \frac{1}{2}(Q_{xx} + Q_{yy} - W) \qquad (14.10)$$

$$m_v = m\sqrt{Q_{vv}} \qquad (14.11)$$

The semiaxes of the error ellipse are m_u and m_v, each drawn in it's respective direction t from point x_a, y_a. If $Q_{xy} = 0$, we obtain from (14.06) $t = 0$ and $90°$; in this case, Q_{xx} and Q_{yy} themselves represent the extreme values. Inversely, when Q_{uu} and Q_{vv} represent the extreme values, then $Q_{uv} = 0$; this could be shown also by a direct demonstration. The error ellipse becomes a circle only when $W = 0$. According to (14.08), this is possible only when $Q_{xx} = Q_{yy}$ and $Q_{xy} = 0$ at the same time. According to (14.06), t is indefinite in this case.

Formulas (14.09) and (14.10) show that Q_{uu} and Q_{vv} can be considered as both solutions of an equation of second degree:

$$\lambda^2 - (Q_{xx} + Q_{yy})\lambda + D = 0 \qquad (14.12)$$

when

$$W^2 = (Q_{xx} \quad Q_{yy})^2 - 4D \qquad (14.13)$$

According to (14.08), we have then

$$D = Q_{xx}Q_{yy} - Q_{xy}^2 = \begin{vmatrix} Q_{xx} & Q_{xy} \\ Q_{xy} & Q_{yy} \end{vmatrix} \qquad (14.14)$$

In other words, D is the determinant of the matrix of the weight coefficients. Equation (14.12) can be written also in form

$$\begin{vmatrix} Q_{xx} - \lambda, & Q_{xy} \\ Q_{xy}, & Q_{yy} - \lambda \end{vmatrix} = 0 \qquad (14.15)$$

In practice, it is not necessary to compute the weight coefficients Q. If we have two parameters only, we can start from the coefficients of the two normal equations. If there are more parameters and we are interested in two only, we arrange these two on the last places. After the elimination of the others by the algorithm of Gauss, we obtain two equations with symmetrical coefficients:

$$[aa]\, x + [ab]\, y = c_x$$
$$[ab]\, x + [bb]\, y = c_y \qquad (14.16)$$

Denoting the determinant of the coefficients

$$d = \begin{vmatrix} [aa] & [ab] \\ [ab] & [bb] \end{vmatrix} \qquad (14.17)$$

we could compute easily the weight coefficients

$$Q_{xx} = \frac{[bb]}{d}$$
$$Q_{yy} = \frac{[aa]}{d} \qquad (14.18)$$
$$Q_{xy} = - \frac{[ab]}{d}$$

and then insert these expressions into (14.07) - (14.10). Eventually, we obtain the following set of computation formulas:

$$w \sin 2t = - 2\, [ab]$$
$$w \cos 2t = [bb] - [aa] \qquad (14.19)$$

$$Q_{uu} = \frac{2}{[aa] + [bb] - w}$$
$$Q_{vv} = \frac{2}{[aa] + [bb] + w} \qquad (14.20)$$

As an example, we take the coefficients of the reduced normal equations given in Table V/9.7:

$$[aa] = 1.47$$
$$[bb] = 6.83$$
$$[ab] = - 1.33$$

and the standard error of the weight unit $m = 0''84$. Formulas (14.19), (14.20), (14.02) and (14.11) give

$w \sin 2t$	=	+ 2.66	$\tan 2t$	=	+ 0.496
$w \cos 2t$	=	+ 5.36	$2t$	=	$26^\circ 4$
w	=	5.98	t	=	$13^\circ 2$
Q_{uu}	=	0.862	m_u	=	0.78 dm
Q_{vv}	=	0.140	m_v	=	0.31 dm

This ellipse is shown in Fig. 9, where an extra feature, m_t, shows how the standard error in an arbitrary direction *t* could be represented. Another numerical example is given in Chapter 16.3.

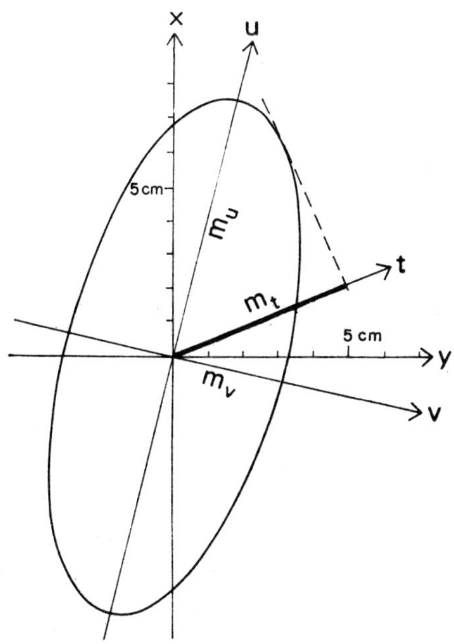

Fig. 9 *Error ellipse*

14.2 *Latent roots and vectors of the weight coefficients*

The discussions of the preceding chapter will be generalized here for the case of more than two parameters. The corresponding "error ellipsoids" in a multidimensional space, however, have no practical merit for more than three parameters. The computation formulas for the three-dimensional error ellipsoids are given in the next chapter but the fundamental theory is given here in a general form because this theory has important applications outside the domain of adjustments.

We shall use a double notation, *viz.* matrices for the general case and the classical symbols for the three-dimensional case in order to make the reading easier. To begin with, the parameters are denoted by vector $\mathbf{x}$ or by components x, y, z. An orthogonal transformation gives a new vector

$$\mathbf{u} = \mathbf{G}^T \mathbf{x} \tag{14.21}$$

or new variables like

$$u = \mathbf{g}_u^T \mathbf{x} \quad \text{or} \quad ax + by + cz \tag{14.22}$$

$$v = \mathbf{g}_v^T \mathbf{x} \quad \text{or} \quad a'x + b'x + c'z$$

The components a, b, c mean the cosines of the angles formed by u - axis with the axes x, y, z. Accordingly, they are tied together by condition

$$\mathbf{g}_u^T \mathbf{g}_u \quad \text{or} \quad a^2 + b^2 + c^2 = 1 \tag{14.23}$$

Similar formulas refer to the other vectors $\mathbf{g}_v$. Because the axes u and v are perpendicular to each other, we have also

$$\mathbf{g}_u^T \mathbf{g}_v \quad \text{or} \quad aa' + bb' + cc' = 0 \tag{14.24}$$

All these conditions can be expressed by one matrix equation

$$\mathbf{G}^T \mathbf{G} = \mathbf{I} \tag{14.25}$$

In other words, the matrix of the orthogonal transformation, representing a rotation in a multidimensional space, has such a property that it's transpose is, at the same time, it's inverse matrix:

$$\mathbf{G}^T = \bar{\mathbf{G}} \tag{14.26}$$

The rotation can be reversed by formula

$$\mathbf{x} = \mathbf{G}\,\mathbf{u} \tag{14.27}$$

The matrix of weight coefficients of the parameters is in our double notation

$$\mathbf{Q}_x \quad \text{or} \quad \begin{Bmatrix} Q_{xx} & Q_{xy} & Q_{xz} \\ Q_{xy} & Q_{yy} & Q_{yz} \\ Q_{xz} & Q_{yz} & Q_{zz} \end{Bmatrix} \tag{14.28}$$

This matrix is supposed to be known. The corresponding matrix for the new variables is according to (12.54)

$$\mathbf{Q}_u = \mathbf{G}^T \mathbf{Q}_x \mathbf{G} \tag{14.29}$$

For one of the new variables we obtain

$$\mathcal{Q}_{uu} = \mathbf{g}_u^T \, Q_x \, \mathbf{g}_u \tag{14.30a}$$

or

$$\begin{aligned} \mathcal{Q}_{uu} = a^2 \mathcal{Q}_{xx} &+ 2ab\, \mathcal{Q}_{xy} + 2ac\, \mathcal{Q}_{xz} \\ &+ b^2\, \mathcal{Q}_{yy} + 2bc\, \mathcal{Q}_{yz} \\ &+ c^2\, \mathcal{Q}_{zz} \end{aligned} \tag{14.30b}$$

Now we shall find a rotation such that $\mathcal{Q}_{uu}$ reaches the greatest possible value. For this purpose, we write δ = maximum where

$$\begin{aligned} \delta &= \tfrac{1}{2} \left\{ \mathcal{Q}_{uu} - \lambda(\mathbf{g}_u^T \, \mathbf{g}_u - 1) \right\} \\ \text{or } &\tfrac{1}{2} \left\{ \mathcal{Q}_{uu} - \lambda(a^2 + b^2 + c^2 - 1) \right\} \end{aligned} \tag{14.31}$$

which, according to (14.23), leads to the same result. Writing all partial derivatives of δ with respect to the components of $\mathbf{g}_u$ or $a, b, c = 0$ we obtain

$$\mathbf{g}_u^T (Q_x - \lambda I) = \mathbf{o} \tag{14.32a}$$

or

$$\begin{aligned} \frac{\partial \delta}{\partial a} &= a(\mathcal{Q}_{xx} - \lambda) + b\, \mathcal{Q}_{xy} + c\, \mathcal{Q}_{xz} = 0 \\ \frac{\partial \delta}{\partial b} &= a\, \mathcal{Q}_{xy} + b(\mathcal{Q}_{yy} - \lambda) + c\, \mathcal{Q}_{yz} = 0 \\ \frac{\partial \delta}{\partial c} &= a\, \mathcal{Q}_{xz} + b\, \mathcal{Q}_{yz} + c(\mathcal{Q}_{zz} - \lambda) = 0 \end{aligned} \tag{14.32b}$$

These equations have a trivial solution $a = b = c = 0$ which, however, does not satisfy the condition (14.23). The other solutions are possible only when the determinant of the coefficients in these equations is zero, cfr. rule 14, Chapter 12.2. In this way, we obtain one equation

$$|Q_x - \lambda I| = 0 \tag{14.33a}$$

or

$$\begin{vmatrix} \mathcal{Q}_{xx} - \lambda & \mathcal{Q}_{xy} & \mathcal{Q}_{xz} \\ \mathcal{Q}_{xy} & \mathcal{Q}_{yy} - \lambda & \mathcal{Q}_{yz} \\ \mathcal{Q}_{xz} & \mathcal{Q}_{yz} & \mathcal{Q}_{zz} - \lambda \end{vmatrix} = 0 \qquad (14.33b)$$

which is an obvious generalization of (14.15).

Accordingly, if vector $\mathbf{x}$ consists of ν components, we should solve an equation of ν:th degree which leads to unhandy computations. The solution gives ν values of λ which are called *latent roots* or *eigenvalues* of matrix $\mathbf{Q}_x$. In adjustments, all these values are positive real numbers.

Each value λ_u is connected to a vector $\mathbf{g}_u$ or to components a, b, c, which can be computed from equations (14.32b). These vectors are called *latent vectors* or *eigenvectors* of $\mathbf{Q}_x$. Because the matrix of coefficients in equations (14.32) is singular, the solution gives first the relative values of the components only; then the condition (14.23) gives the components themselves.

Inserting the equations (14.32b) into (14.30b) we see that anyone of the latent roots λ_u is identical with one of the extreme values $\mathcal{Q}_{uu}$ from which one is an absolute maximum, another an absolute minimum. The third solution represents either a maximum or a minimum according to the plane in which the respective axis is rotated through the position represented by the direction of vector $\mathbf{g}_u$.

In the general case (14.32a) is replaced by equation

$$\mathbf{Q}_x \, \mathbf{g}_u = \lambda_u \, \mathbf{g}_u \qquad (14.34)$$

which is inserted into (14.30a) to give by the use of (14.23)

$$\mathcal{Q}_{uu} = \lambda_u \, \mathbf{g}_u^T \, \mathbf{g}_u = \lambda_u \qquad (14.35)$$

The cofactor of components u and v is according to (14.29)

$$\mathcal{Q}_{uv} = \mathbf{g}_u^T \, \mathbf{Q}_x \, \mathbf{g}_v \qquad (14.36)$$

which by the aid of (14.32a) and (14.24) gives

$$\mathcal{Q}_{uv} = \lambda_u \, \mathbf{g}_u^T \, \mathbf{g}_v = 0 \qquad (14.37)$$

Thus the new parameters are uncorrelated in all combinations.

14.3 *Error ellipsoid*

Error ellipsoid of a set of values **x** or x, y, z is the multidimensional "ellipsoid" the center of which has the components of **x** as coordinates and the semiaxes of which are

$$m_u = m\sqrt{Q_{uu}} = m\sqrt{\lambda_u} \tag{14.38}$$

each drawn in the direction of the respective vector $\mathbf{g}_u$.

We suppose first that **x** consists of the three last parameters of an adjustment by method *A*. When all the other parameters have been eliminated from the normal equations by Gauss' algorithm, the coefficients of these three parameters form a symmetrical matrix

$$\mathbf{N} = \begin{bmatrix} n_{xx} & n_{xy} & n_{xz} \\ n_{xy} & n_{yy} & n_{yz} \\ n_{xz} & n_{yz} & n_{zz} \end{bmatrix} \tag{14.39}$$

The determinant of **N** could be computed by formula

$$\begin{aligned} d = |\mathbf{N}| = \; & n_{xx}n_{yy}n_{zz} + 2n_{xy}n_{xz}n_{yz} \\ & - n_{xx}n_{yz}^2 - n_{yy}n_{xz}^2 - n_{zz}n_{xy}^2 \end{aligned} \tag{14.40}$$

As an alternative, we compute the minor determinants

$$\begin{aligned} d_{xx} &= n_{yy}n_{zz} - n_{yz}^2 \\ d_{yy} &= n_{xx}n_{zz} - n_{xz}^2 \\ d_{zz} &= n_{xx}n_{yy} - n_{xy}^2 \\ d_{xy} &= n_{xz}n_{yz} - n_{zz}n_{xy} \\ d_{xz} &= n_{xy}n_{yz} - n_{yy}n_{xz} \\ d_{yz} &= n_{xy}n_{xz} - n_{xx}n_{yz} \end{aligned} \tag{14.41}$$

by the aid of which we obtain the determinant in three ways:

$$\begin{aligned} d &= n_{xx}d_{xx} + n_{xy}d_{xy} + n_{xz}d_{xz} \\ &= n_{xy}d_{xy} + n_{yy}d_{yy} + n_{yz}d_{yz} \\ &= n_{xz}d_{xz} + n_{yz}d_{yz} + n_{zz}d_{zz} \end{aligned} \tag{14.42}$$

From the minor determinants, we obtain the weight coefficients dividing them by d or multiplying them by

$$\mathcal{D} = |\mathbf{Q}| = \frac{1}{d} \tag{14.43}$$

Then we have

$$Q_{ij} = \mathcal{D}\, d_{ij} \tag{14.44}$$

If $\mathbf{x}$ does not mean the parameters of an adjustment, or if the adjustment has been carried out by method $\mathcal{B}$, we must start from the weight coefficients $\mathbf{Q}$. First we compute the minor determinants

$$\begin{aligned} \mathcal{D}_{xx} &= Q_{yy}Q_{zz} - Q_{yz}^{2} \\ \mathcal{D}_{xy} &= Q_{xz}Q_{yz} - Q_{zz}Q_{xy} \end{aligned} \tag{14.45}$$

etc. Then we compute the determinant

$$\mathcal{D} = Q_{xx}\mathcal{D}_{xx} + Q_{xy}\mathcal{D}_{xy} + Q_{xz}\mathcal{D}_{xz} \tag{14.46}$$

In both cases, we need the coefficients

$$\begin{aligned} q &= \frac{\mathcal{D}}{3}(d_{xx} + d_{yy} + d_{zz}) = \frac{1}{3}(Q_{xx} + Q_{yy} + Q_{zz}) \\ n &= \frac{\mathcal{D}}{3}(n_{xx} + n_{yy} + n_{zz}) = \frac{1}{3}(\mathcal{D}_{xx} + \mathcal{D}_{yy} + \mathcal{D}_{zz}) \end{aligned} \tag{14.47}$$

by the aid of which we can write the equation (14.33b) in form

$$\lambda^{3} - 3q\lambda^{2} + 3n\lambda - \mathcal{D} = 0 \tag{14.48}$$

The solution of such an equation of third degree can be found by formulas

$$\begin{aligned} k &= \sqrt{q^{2} - n} \\ \cos\phi &= \frac{1}{2k^{3}}(\mathcal{D} + 2q^{3} - 3qn) \\ \lambda_1 &= q + 2k\cos\frac{\phi}{3} \end{aligned} \tag{14.49}$$

First, k is taken with the positive sign and ϕ between 0° and 180°. When λ_1 has been computed, we repeat the computation for $\phi + 360^\circ$ and $\phi - 360^\circ$. In this way, we obtain λ_2 and λ_3. As an alternative

$$\begin{aligned} \lambda_2 &= q - k\left(\cos\frac{\phi}{3} + \sqrt{3}\sin\frac{\phi}{3}\right) \\ \lambda_3 &= q - k\left(\cos\frac{\phi}{3} - \sqrt{3}\sin\frac{\phi}{3}\right) \end{aligned} \tag{14.50}$$

For each value of λ, we compute the components of the corresponding latent vector by either one of these alternative sets of formulas:

$$\begin{aligned} \alpha &= n_{yz} + \lambda d_{yz} \\ \beta &= n_{xz} + \lambda\, d_{xz} \\ \gamma &= n_{xy} + \lambda d_{xy} \\ g &= \sqrt{\frac{1}{\alpha^2} + \frac{1}{\beta^2} + \frac{1}{\gamma^2}} \qquad (14.51) \\ a &= 1/g\alpha \\ b &= 1/g\beta \\ c &= 1/g\gamma \end{aligned}$$

or

$$\begin{aligned} A &= \mathcal{D}_{yz} + \lambda \mathcal{Q}_{yz} \\ B &= \mathcal{D}_{xz} + \lambda \mathcal{Q}_{xz} \\ C &= \mathcal{D}_{xy} + \lambda \mathcal{Q}_{xy} \\ G &= \sqrt{\frac{1}{A^2} + \frac{1}{B^2} + \frac{1}{C^2}} \qquad (14.52) \\ a &= 1/GA \\ b &= 1/GB \\ c &= 1/GC \end{aligned}$$

The automatic computers usually have ready programs for the computation of the greatest latent root with the corresponding latent vector. The computation method is based on an iteration according to (14.34). First, vector $\mathbf{g}$ is replaced by any vector satisfying (14.23), *e.g.* one of the components = 1, all the others = 0. This vector $\mathbf{g}_o$ is multiplied several times by matrix $\mathbf{Q}_x$: $\mathbf{g}_1 = \mathbf{Q}_x\mathbf{g}_o$, $\mathbf{g}_2 = \mathbf{Q}_x\mathbf{g}_1 \ldots$, $\mathbf{g}_i = \mathbf{Q}_x\mathbf{g}_{i-1}$. After each multiplication the factor λ_i is written down, showing the proportion in which greatest of the components has been changed. After repeated multiplications, this proportion tends to a limit which is the greatest latent root of $\mathbf{Q}_x$ while the components of $\mathbf{g}_i$ tend to the values which are proportional to the components of the corresponding latent root. A similar treatment of matrix $\bar{\mathbf{Q}} = \mathbf{N}$ gives the smallest latent root of $\mathbf{Q}$ and the corresponding latent vector. The iteration proceeds very slowly if $\mathbf{g}_o$ happens to be close to a wrong latent vector or if the two greatest latent roots are almost equal.

As an example, we compute the error ellipsoid for point 13 which appears in Table II/17.5. The weight coefficients are

$$\mathbf{Q} = \begin{Bmatrix} 0.927 & 0.173 & 0.248 \\ 0.173 & 2.301 & 1.515 \\ 0.248 & 1.515 & 2.296 \end{Bmatrix}$$

The minor determinants according to (14.45) are

$$\begin{Bmatrix} 2.985 & -0.020 & -0.310 \\ -0.020 & 2.066 & -1.362 \\ -0.310 & -1.362 & 2.102 \end{Bmatrix}$$

D = 2.686, according to (14.46). Further, q = 1.841 and n = 2.384, according to (14.47). Equation (14.48) gives

$$\lambda^3 - 5.523\,\lambda^2 + 7.152\,\lambda - 2.686 = 0$$

According to (14.49), we have first k = 1.0025, cos ϕ = 0.9909 and $\phi = 7^\circ 44'$. The rest of the solution runs as follows:

$\frac{\phi}{3}$	$2^\circ 35'$	$122^\circ 35'$	$242^\circ 35'$
$\cos\frac{\phi}{3}$	0.999	- 0.538	- 0.461
λ	3.84	0.76	0.92
A	4.456	- 0.211	0.032
B	0.642	- 0.122	- 0.082
C	0.644	0.111	0.139
G	2.211	13.07	34.31
a	0.101	- 0.363	0.911
b	0.705	- 0.627	- 0.356
c	0.702	0.689	0.210

The iteration method gives

g_0	g_1	g_2	g_3	g_4	g_5	
0	0.248	1.061	4.09	15.65	60.0	a = 0.101
0	1.515	7.007	27.86	107.82	415.0	b = 0.703
1	2.296	7.628	28.39	108.41	416.1	c = 0.704
λ	2.296	3.322	3.722	3.818	3.839	

This error ellipsoid has one longer axis and two shorter ones. The components of the latent vector show that the direction of the long axis is about halfway between $+y$ - and $+z$ - axes, with a slight slanting towards $+x$ - axis.

15 *SPECIAL METHODS*

15.1 *Solution of normal equations by successive approximations*

Iteration means a method of computation by successive steps, *e.g.* for the prediction of the orbit of a missile. Often this name is used also for the method of successive approximations in which case "trial and error" would be a better description of the procedure.

The solution of normal equations by an iteration could be described theoretically by the following matrix formulas. Let the system of equations to be solved be denoted by

$$\mathbf{N}\,\mathbf{x} + \mathbf{u} = \mathbf{o} \tag{15.01}$$

in which $\mathbf{N}$ is a square matrix, not necessarily symmetrical. Let $\mathbf{x}_1$ be any approximate solution, based on a good guess, a rapid estimation, or a programmed short computation. The correct solution can be denoted by

$$\mathbf{x} = \mathbf{x}_1 + \mathbf{z}_1 \tag{15.02}$$

where $\mathbf{z}_1$ is a new unknown vector. Inserting (15.02) into (15.01), we obtain

$$\mathbf{N}\,\mathbf{z}_1 + \mathbf{u}_1 = \mathbf{o} \tag{15.03}$$

where

$$\mathbf{u}_1 = \mathbf{u} + \mathbf{N}\,\mathbf{x}_1 \tag{15.04}$$

is a new known vector but the coefficients $\mathbf{N}$ are the same as in (15.01).

The "good guess", estimation or the short programmed computation is repeated several times. Generally, we have equations

$$\mathbf{N}\,\mathbf{z}_i + \mathbf{u}_i = \mathbf{o} \tag{15.05}$$

for which we need only the approximate solution $\mathbf{x}_{i+1}$. Then

$$\mathbf{z}_i = \mathbf{x}_{i+1} + \mathbf{z}_{i+1} \tag{15.06}$$

defines the new unknown vectors $\mathbf{z}_{i+1}$ and

$$\mathbf{u}_{i+1} = \mathbf{u}_i + N\,\mathbf{x}_{i+1} \qquad (15.07)$$

the new known vectors. The only requirement for a rapid iteration is that, in average, the elements of $\mathbf{u}_{i+1}$ are at each step clearly smaller than those of $\mathbf{u}_i$. The definitive solution is

$$\mathbf{x} = \mathbf{x}_1 + \mathbf{x}_2 + \mathbf{x}_3 + \ \ldots. \qquad (15.08)$$

The most important detail of the general procedure described above is, of course, the selection of the approximate solutions $\mathbf{x}_i$. The easiest way is to deal each time with one component only. For this purpose, that equation is picked which has the greatest component of $\mathbf{u}_i$. Denoting this component by $u_{i,m}$, we divide it by the diagonal element n_{mm} obtaining a correction

$$x_{i+1,m} = -\frac{u_{i,m}}{n_{mm}} \qquad (15.09)$$

for the corresponding unknown. Theoretically, it would be still better to pick that equation for which the value of $(u_{i,m})^2/n_{mm}$ is greatest but, in practice, we should avoid such systems of equations in which the diagonal coefficients n_{mm} have very unequal values.

In spite of formula (15.09), the division should not be performed exactly. It is quite sufficient — and for the lightening of the later computations even profitable — to use rounded off values like

10, 5, 2, 1, 0.5, 0.2 *etc.*

with proper signs. Division (15.09) and the multiplications implied by (15.07) can then be performed entirely by mental operations. In fact, the most peculier feature of the iteration method is that the equations (15.01) can be solved without any computing devices.

In order to secure equations in which the diagonal coefficients are of the same order, the units of the unknown parameters should be chosen carefully or the condition equations multiplied by suitable constants. If this has been omitted, some of the normal equations may have a diagonal coefficient exceptionally large or small. In such a case, the diagonal coefficient can be divided or multiplied by 4, 25 or 100 when, at the same time, the other terms of the equation are divided or multiplied by 2, 5 or 10, respectively.

For the practical arrangement of the manual computations, we recommend the scheme given in Table I. After the table showing the coefficients of the normal equations, write the table in which each row represents the successive vectors $\mathbf{u}_i^T$. An extra column is reserved for the sum of the elements of each row. The last column is reserved for the estimated elements $x_{i,m}$.

At each step of the iteration, the greatest element of the current vector $\mathbf{u}_i^T$ is picked up, underlined and divided by that diagonal coefficient n_{mm} which is in the same column. The rounded-off value 5, 2 or 1 with changed sign and in correct units is written into the last column. All coefficients of the used equation are multiplied by this constant and the *exact* values of the products are added to the respective elements of $\mathbf{u}_i^T$. In this way the next row $\mathbf{u}_{i+1}^T$ is obtained.

When picking up the greatest element u_i, there usually are several of those which would lead to an equal correction $x_{i,m}$. In such a case, it is advisable to observe the coefficients which express the correlations between the greatest elements and to postpone the picking of those which would anyway be reduced by the dealing with the others; in other words, start from those which increase the other elements. A strong rounding-off upwards is necessary, sometimes a double or triple value of the exact quotient indicated by (15.09) is found to be best. If the system of normal equations is suitable for the solution by iterations, each of the values 5, 2, and 1 should appear not more than once for the same unknown. If it turns out that the same correction appears several times for the same unknown, it is just a waste of time to apply the iteration method to the single rows one by one.

When the corrections are not more than half of the last desired unit, the iteration is interrupted. The corrections $x_{i,m}$ are summed up following the unterlined elements in each column m.

Table I *Normal equations and the solution by successive approximations*

613 x	y	614 x	y	615 x	y	Σ	ℓ
6.00	-1.01	+0.06	-0.19	-1.12	+0.57	+4.31	+1.44
	9.35	- 5	- 88	+1.07	-1.23	+7.25	-5.76
		4.99	+ 86	-1.57	- 31	+3.98	+ 64
			4.97	-1.11	-1.64	+2.01	+2.55
				6.62	- 10	+3.79	-6.54
					6.80	+4.09	-1.09
							-8.76

ℓ						Σ	Δx
+1.44	-5.76	+0.64	+2.55	-6.54	-1.09	-8.76	+1
+ 32	-4.69	- 93	+1.44	+ 8	-1.19	-4.97	+0.5
- 185	- 15	- 955	+1.000	+ 615	-1.805	-1.345	+0.2
- 71	- 261	-1.017	+ 672	+ 595	- 445	- 527	+0.2
- 59	- 271	- 19	+ 844	+ 281	- 507	+ 269	-0.2
- 21	- 95	- 191	- 150	+ 503	- 179	- 133	-0.1
- 91	- 202	- 34	- 039	- 159	- 169	- 512	+0.03
+ 108	- 239	- 43	- 88	- 162	+ 35	- 389	+0.02
+ 88	- 52	- 44	- 106	- 141	+ 10	- 245	+0.02
+ 66	- 31	- 75	- 128	- 9	+ 8	- 169	+0.02
- 62	- 48	- 58	- 29	- 31	- 25	- 129	+0.01
+ 63	- 49	- 8	- 20	- 47	- 28	- 89	-0.01
+ 3	- 39	- 9	- 18	- 36	- 34	- 133	+0.005
- 3	- 34	- 17	- 24	- 3	- 34	- 115	+0.005
0	- 40	- 19	- 32	- 4	0	- 95	+0.01
- 2	- 49	- 10	+ 18	- 15	- 16	- 74	+0.005
- 7	- 2	- 10	+ 14	- 10	- 22	- 38	+0.005
- 4	- 8	- 12	+ 6	- 10	+ 12	- 18	
Σ Δ x.							
- 0.01	+ 0.53	+0.21	-0.17	+0.93	+0.24	decimeters	
Check							
- 0.043	+ 3.842	+0.836	-0.342	+3.524	+0.982	+8.799	
					pro	8.76	

As an example, the normal equations given in Table VII/9.6 are solved again in Table I using the method described above.

There are many variations of this elementary method, mostly devised for the speeding up of the convergence. Most of the special tricks, however, hardly can be adapted for the programming of the automatic computers. The iteration is not at all suitable when some observation equations contain many parameters or when some observations are connected by many conditions. Inversely, the iteration is very advantageous when the observations refer mostly to the differences of parameters or when the conditions are formed around small loops of the observation net. In such cases, the normal equations often are omitted and the instructions for an iteration method are given in such a form that the computations can be carried out by a person who has no learning in adjustments. Best known are the method of successive mean values of parameters and the successive elimination of misclosures from the original conditions. An investigation of the instructions reveals that the coefficients of the normal equations, anyway, must be computed in several phases of these procedures.

Another disadvantage of the iterative methods is that the weight coefficients are not obtained without long additional computations. There is, however, a possibility to unite the advantages of the iterative methods with those of the Gauss' algorithm. An experienced person who is able to carry out mental computations can find easily the approximative values of elements of Cholesky's matrix **K**; the accuracy of one significant figure is quite sufficient. By the aid of this matrix, the approximate elements e_i and x_i can be computed; see (12.36) and (12.39). The latter elements are rounded off and new elements u_{i+1} must be computed with a full precision according to (15.07). The matrix **K** obtained during the first step can be used again each time for the later steps.

15.2 *Solution of normal equations by groups*

The number of normal equations in an adjustment is often several hundreds; the high speed computers can deal with thousands of simultaneous linear equations. The practical solution of such great systems is made possible by the partitioning of the matrices involved. The partitioned matrices are dealt with one by one and the results are stored outside the computer on punched cards or tapes. The same idea can be applied by manual computations writing down intermediate results in form of matrices **Z** and vectors **z** which will be explained below. These "Zwischenkorrelaten" were introduced in the adjustments by *Boltz* before matrix notations were generally known.

The general idea of the partitioning can be seen from the following scheme where the normal equations have been divided into three groups:

$$\begin{aligned} \mathbf{N}_1\mathbf{x}_1 \qquad\qquad\quad\; \mathbf{N}_{13}\mathbf{x}_3 + \mathbf{u}_1 &= \mathbf{o} \\ \mathbf{N}_2\mathbf{x}_2 + \mathbf{N}_{23}\mathbf{x}_3 + \mathbf{u}_2 &= \mathbf{o} \\ \mathbf{N}_{13}^T\mathbf{x}_1 + \mathbf{N}_{23}^T\mathbf{x}_2 + \mathbf{N}_3\mathbf{x}_3 + \mathbf{u}_3 &= \mathbf{o} \end{aligned} \qquad (15.10)$$

In geodetic applications, it is always possible to arrange the normal equations in such an order that one group of coefficients, corresponding to the missing matrix $\mathbf{N}_{12}$ in the scheme above, consists of a zero matrix, *e.g.* taking first the both ends of a longish net and then the middle part of it.

Suppose that there are ready programs for the available computer which gives the inverse matrices $\bar{\mathbf{N}}_1$ and $\bar{\mathbf{N}}_2$. Compute then

$$\begin{aligned} \mathbf{Z}_{13} &= \bar{\mathbf{N}}_1\mathbf{N}_{13} \\ \mathbf{Z}_{23} &= \bar{\mathbf{N}}_2\mathbf{N}_{23} \end{aligned} \qquad (15.11)$$

$$\begin{aligned} \mathbf{Z}_3 &= \mathbf{N}_3 - \mathbf{Z}_{13}^T\mathbf{N}_{13} - \mathbf{Z}_{23}^T\mathbf{N}_{23} \\ \mathbf{z} &= \mathbf{u}_3 - \mathbf{Z}_{13}^T\mathbf{u}_1 - \mathbf{Z}_{23}^T\mathbf{u}_2 \end{aligned} \qquad (15.12)$$

The final solution is now

$$\begin{aligned} \mathbf{x}_3 &= -\bar{\mathbf{Z}}_3\mathbf{z} \\ \mathbf{x}_2 &= -\bar{\mathbf{N}}_2\mathbf{u}_2 - \mathbf{Z}_{23}\mathbf{x}_3 \\ \mathbf{x}_1 &= -\bar{\mathbf{N}}_1\mathbf{u}_1 - \mathbf{Z}_{13}\mathbf{x}_3 \end{aligned} \qquad (15.13)$$

In manual computations, the inverse matrices are not necessary. Instead, the partitioned matrices **R** and **T** with some intermediate forms are required. The solution in groups then is nothing more than an arrangement of the computations according to Gauss' algorithm on many separated sheets of paper with additional pages for the intermediate results.

The inverse matrix of the complete matrix **N** can be obtained from the partitioned matrices of various groups as follows:

$$\bar{N} = \begin{bmatrix} \bar{N}_1 + (Z_{13}\bar{Z}_3)\, Z_{13}^T & (Z_{13}\bar{Z}_3)Z_{23}^T; & -\, Z_{13}\bar{Z}_3 \\ Z_{23}(\bar{Z}_3 Z_{13}^T) & \bar{N}_2 + (Z_{23}\bar{Z}_3)Z_{23}^T; & -\, Z_{23}\bar{Z}_3 \\ -\bar{Z}_3 Z_{13}^T & -\bar{Z}_3 Z_{23}^T & \bar{Z}_3 \end{bmatrix} \qquad (15.14)$$

15.3 *The general problem of adjustments*

So far, we have considered the methods A and B or adjustments with observation equations and those with condition equations, as two separate problems but here they will taken as special cases of one general problem with initial equations of type

$$\mathbf{A}\,\mathbf{x} + \mathbf{B}^T\,\mathbf{v} + \mathbf{w} = \mathbf{o} \qquad (15.15)$$

If $\mathbf{B} = -\mathbf{I}$, (15.15) gives observation equations, if $\mathbf{A} = \mathbf{0}$, it gives condition equations.

Equations of type (15.15) appear if from the original observation equations some parameters are eliminated. Then vector $\mathbf{x}$ represents the remaining parameters and the matrix $\mathbf{B}$ describes the method of elimination. In practice, this case occurs when the values of some parameters have been immediately observed. Then these parameters are no more elements of vector $\mathbf{x}$ but additional elements of vector $\mathbf{l}$ with unknown residuals $\mathbf{v}$. As an example, consider a photogrammetric triangulation in which the geodetic coordinates of control points are not accurate enough but require corrections with due weights.

The solution of (15.15) can be found by the principle of least squares if the minimum value of expression

$$2s = \mathbf{v}^T\,\mathbf{P}\,\mathbf{v} - 2\,\mathbf{k}^T\,(\mathbf{A}\mathbf{x} + \mathbf{B}^T\,\mathbf{v} + \mathbf{w}) \qquad (15.16)$$

is determined. In this expression, $\mathbf{x}$ and $\mathbf{v}$ are variables, $\mathbf{k}$ means unknown constants. The partial derivatives of (15.16) with respect to the variables must be zero:

$$\frac{\partial s}{\partial \mathbf{v}} = \mathrm{P}\,\mathbf{v} - \mathrm{B}\,\mathrm{k} = \mathbf{o} \tag{15.17}$$

$$\frac{\partial s}{\partial \mathbf{x}} = \mathbf{A}^{\mathrm{T}}\mathrm{k} = \mathrm{o} \tag{15.18}$$

(15.17) gives

$$\mathbf{v} = \bar{\mathbf{P}}\,\mathrm{B}\,\mathbf{k} \tag{15.19}$$

Inserting this into (15.15) and denoting

$$\mathrm{M} = \mathrm{B}^{\mathrm{T}}\bar{\mathrm{P}}\,\mathrm{B} \tag{15.20}$$

the following group of equations is obtained:

$$\mathrm{M}\,\mathrm{k} + \mathbf{A}\,\mathbf{x} + \mathbf{w} = \mathrm{o} \tag{15.21}$$

Together with (15.18), these equations form a symmetrical group which could be solved as usual normal equations with respect to the unknown elements of **k** and **x**. The equations can be solved also by groups. Compute first

$$\mathrm{Z} = \bar{\mathbf{M}}\,\mathbf{A} \tag{15.22}$$

$$\mathbf{N} = \mathrm{Z}^{\mathrm{T}}\,\mathbf{A} \tag{15.23}$$

$$\mathbf{u} = \mathrm{Z}^{\mathrm{T}}\,\mathbf{w} \tag{15.24}$$

Then a matrix equation is obtained:

$$\mathbf{N}\,\mathbf{x} + \mathbf{u} = \mathrm{o} \tag{15.25}$$

the solution of which is

$$\mathbf{x} = -\,\bar{\mathbf{N}}\,\mathbf{u} \tag{15.26}$$

$$\mathbf{k} = -\,\bar{\mathrm{M}}\,\mathbf{w} - \mathrm{Z}\,\mathbf{x} \tag{15.27}$$

The residuals **v** can be computed by (15.19) and checked by

$$[pvv] = -\,[wk] \tag{15.28}$$

The variance of unit weight can be estimated by

$$m^2 = \frac{[pvv]}{r - u} \tag{15.29}$$

where r is the number of elements w and u the number of elements x.

15.4 *Parameters with conditions*

Observation equations with conditions between the parameters can be obtained in practice in two ways:

1. There are more parameters than the minimum number required by the theory wherefore conditions between the parameters are necessart.
2. Some observations are regarded errorless, in other words, the corresponding observation equations with infinite weights contain no residuals.

In both cases, there are two matrix equations:

$$\mathbf{A}\ \mathbf{x} + \mathbf{l}_1 = \mathbf{v} \quad \text{with weights } \mathbf{P} \tag{15.30}$$

$$\mathbf{B}^t\mathbf{x} + \mathbf{l}_2 = \mathbf{o} \tag{15.31}$$

The solution could be found by elimination of as many parameters as there are conditions. Another method of solution can be found in the following way. The minimum value of expression

$$2s = \mathbf{v}^T\ \mathbf{P}\ \mathbf{v} + 2\ \mathbf{k}^T\ (\mathbf{B}^T\mathbf{k} + \mathbf{l}_2) \tag{15.32}$$

is determined. Denoting

$$\mathbf{N} = \mathbf{A}^T\ \mathbf{P}\ \mathbf{A} \tag{15.33}$$

$$\mathbf{u} = \mathbf{A}^T\ \mathbf{P}\ \mathbf{l}_1 \tag{15.34}$$

we can write

$$\frac{\partial s}{\partial \mathbf{x}} = \mathbf{N}\ \mathbf{x} + \mathbf{B}\ \mathbf{k} + \mathbf{u} = \mathbf{o} \tag{15.35}$$

Together with (15.31), these equations form a symmetrical system which could be solved as usual normal equations. The equations can be solved also by groups. Compute first

$$\mathbf{Z} = \bar{\mathbf{N}}\ \mathbf{B} \tag{15.36}$$

$$\mathbf{M} = \mathbf{Z}^T\ \mathbf{B} \tag{15.37}$$

$$\mathbf{w} = \mathbf{Z}^T\ \mathbf{u} - \mathbf{l}_2 \tag{15.38}$$

Then a matrix equation is obtained:

$$\mathbf{M}\ \mathbf{k} + \mathbf{w} = \mathbf{o} \tag{15.39}$$

the solution of which is

$$\mathbf{k} = -\bar{\mathbf{M}}\,\mathbf{w} \tag{15.40}$$

$$\mathbf{x} = -\bar{\mathbf{N}}\,\mathbf{u} - \mathbf{Z}\,\mathbf{k} \tag{15.41}$$

The residuals $\mathbf{v}$ can be computed by (15.30) and checked by

$$[pvv] = [p\ell_1\ell_1] + [xu] + [k\ell_2] \tag{15.42}$$

The variance of unit weight can be estimated by

$$m^2 = \frac{[pvv]}{n - u + r} \tag{15.43}$$

where n is the number of elements ℓ_1, r the number of elements ℓ_2 and u the number of elements x.

15.5 *Fictitious observations*

A great number of observed quantities $\mathbf{l}_b$ can often be replaced by a small number of their functions $\mathbf{y}_b$ which can be used as fictious observations for an adjustment. The partial derivatives of these functions form a rectangular matrix $\mathbf{F}$ by the aid of which we can write.

$$d\mathbf{y} = \mathbf{F}^T\,d\mathbf{l} \tag{15.44}$$

The weight coefficients of $\mathbf{y}_b$ can be found from matrix product

$$\mathbf{Q} = \mathbf{F}^T\,\bar{\mathbf{P}}\,\mathbf{F} \tag{15.45}$$

when $\mathbf{P}$ means the weight matrix of the original observations. Usually $\mathbf{P}$ and $\bar{\mathbf{P}}$ are diagonal matrices but $\mathbf{Q}$, anyway, is a symmetrical square matrix.

Accordingly, when the original observed quantities $\mathbf{l}_b$ or a part of them are replaced by these fictitious observations, the latter must be considered as correlated. Anyway, the matrix formulas of the adjustment are valid in the case of correlated observations. In method A, the observation equations are formed for the fictious observations using the inverse matrix $\bar{\mathbf{Q}}$ instead of $\mathbf{P}$. In method B, the misclosures must be expressed as functions of $\mathbf{y}_b$ and, when the normal equations are formed, the original matrix $\mathbf{Q}$ is used instead of $\bar{\mathbf{P}}$.

In most cases, an original observed quantity already is a fictitious observation, *viz.* the weighted mean of many individual observations or "readings", with the sum of original weights as the final weight. As more complicated examples, three geodetic applications can be mentioned. *Station adjustments* with incomplete sets give fictitious observations with correlations as explained in the next chapter. In Chapter 16.7 we shall discuss the rigorous adjustment of *traverses* replacing all sides and angles of a single chain which leads from one fork point to another by three fictitious observed quantities, *viz.* differences of x- and y-coordinates and of the zero directions at the end points.

In stellar triangulations, a satellite can be photographed simultaneously from two ground stations against the background of stars. An adjustment of all satellite positions with have been observed from both stations gives the Greenwich hour angle and the declination of the line between these stations. These two correlated fictitious observations can be used in the final net adjustment of many ground stations.

An adjustment by fictitious observations is similar to an adjustment by groups. The essential advantage of these methods is that a large net which may overrun the capacity of an automatic computer is divided into small parts for which there are routine programs.

16 GEODETIC APPLICATIONS

16.1 *Station adjustments*

If the horizontal angles at one station have been measured in complete sets the station adjustment is carried out as explained Chapter 4.3.

If the horizontal angles have been observed one by one in all combinations of two directions the station adjustment can be carried out as explained in Chapter 5.5.

In both on these cases, the final values of directions are uncorrelated and equally weighted. The general case where incomplete sets occur is very rare because it occurs in practice, only when an original plan of observations has accidentally failed. A few exceptional stations with incomplete sets could be dealt with in connection with the net adjustment, using proper weights and additional zero parameters for the various groups of sets. By the aid of a preliminary adjustment, however, all original observations at one station can be replaced by one set of fictitious correlated observations. Because the general algebraic formulas of this method are rather complicated, the procedure will be explained here with the aid of a numerical example.

At station P, the directions A, B, C, D, E have been observed in five different combinations, see the first columns of Table I. The weight of each "reading" is supposed to be = 1. Using A as the initial direction, the "shifted" values for the other directions have been taken from sets 1 and 4. In other sets, the readings have been shifted on the ground of these approximate values. The differences approximate minus shifted, denoted by ℓ in the table, are computed and written as the absolute members in observation equations; the coefficients of these equations - merely numbers 1 - are shown in the table too. Then the normal equations are formed in the usual way.

Table I *Station adjustment with incomplete sets*

Set Point	Directions Observed	Shifted	Observation equations 1	2	3	4	5	B	C	D	E	l	v'
1 *A*	0,00000	0,00000	1									0	−44,7
B	23,19293	23,19293	1					1				0	6,0
C	83,63127	83,63127	1						1			0	38,7
2 *B*	0,00000	23,19293		1				1				0	− 6,0
E	170,61071	193,80364		1							1	+ 23	6,0
3 *A*	0,00000	0,00000			1							0	20,0
C	83,63269	83,63269			1				1			−142	−38,7
E	193,80428	193,80428			1						1	− 41	18,7
4 *A*	0,00000	0,00000				1						0	24,6
D	112,10243	112,10243				1				1		0	−88,9
E	193,80387	193,80387				1					1	0	64,3
5 *D*	0,00000	112,10243					1			1		0	88,9
E	81,70475	193,80718					1				1	−331	−88,9
	Approximate		Normal equations									u	
A	0,00000		3					1	1			0	
B	23,19293			2				1			1	+ 23	
C	83,63127				3				1		1	−183	
D	112,10243					3				1	1	0	
E	193,80387						2			1	1	−331	
								2				0	
									2			−142	
										2		0	
											4	−349	
												131935	

Table II

Weight matrix

B	*C*	*D*	*E*	u'
1,17	−0,33		−0,50	− 11,5
	1,33		−0,33	− 81,0
		1,17	−0,83	+165,6
			2,33	−134,0
				65727

Table III

Inverse matrix and solution

B	*C*	*D*	*E*	x
1,14	0,39	0,29	0,40	+ 50,6
	0,92	0,20	0,29	+ 83,3
		1,24	0,53	−113,6
			0,75	+ 39,6
			$[v'v']$	34293

Table IV *Fictitious directions*

P	Adjusted	Weight coefficients					Weight matrix				
A	^{g}0,00000	0,29					3,45				
B	23,19344		0,85	0,10		0,11		1,24	–,18	+,07	–,35
C	83,63210			0,63	–,09				1,63	+,15	–,03
D	112,10129				0,95	0,24				1,24	–,68
E	193,80468					0,46					2,63

For the solution of the normal equations, the orientation parameters of the different sets are eliminated first. The reduced coefficients of variations of the remaining parameters B, C, D, E can be seen in Table II. The inverse matrix and the solution are shown in Table III. The elements of the inverse matrix give the weight coefficients of angles APB, APC, APD and APE. The adjusted values of these angles could be taken in the eventual net adjustment as fictitious observations. However, it is usual to replace these angles by a fictitious set of directions as given in Table IV. The weight coefficients of these directions could be computed in many ways. Here, the inverse weight of direction A is chosen to be = 0.29 and the other weight coefficients = 0. The coefficients of the remaining directions are then obtained from those of the corresponding angles by subtracting the constant 0.29. If the net adjustment is carried out by variation of parameters (and not by conditions), the matrix of weight coefficients must be inverted, see "Weight matrix" of Table IV.

The standard error of unit weight is $m = \sqrt{343/4} = \pm 9''.3$. This is the precision or the internal accuracy of one "reading". After a net adjustment, another standard error can be obtained, which usually is larger than m obtained above; this external accuracy is, of course, more reliable.

Exercise: Prove that if there are n complete sets of s directions and n' additional sets with *one* direction missing, all sets can be replaced by one *uncorrelated* set using the value of readings and weight $n + n'$ for other directions but a corrected value and a reduced weight for the direction with incomplete observations.

If the other directions are D_i $(i = 1, 2 \ldots s - 1)$ and the direction with incomplete observations D_s, the mean values L_i (n complete sets) and L_i' (n' incomplete sets) and C the mean value of all differences $L_i' - L_i$, the corrected value of direction D_s is $L_s + n'C/(n + n')$ and it's weight

$$\frac{n(n + n')(s - 1)}{n(s - 1) + n's}$$

16.2 *Base extension nets*

Base extension nets are in ideal cases rhombuses but, in practice, quadrilaterals with two diagonals. The shorter diagonal is measured, the longer diagonal consists of a side of the main net or of the shorter diagonal of another quadrilateral. In a state triangulation, the base line, or the shorter diagonal of the first quadrilateral, is 4 - 6 km long, the sides of the main net 20 - 40 km long. Because the angles of the quadrilateral should not be smaller than 40°, the extension nets consist usually of two quadrilaterals with a common diagonal 10 - 15 km long, see Fig. 10.

The extension nets are adjusted separately before the adjustment of the main net. As an example, we take the extension net of Lappträsk base in Finland. Table I gives the list of the observed directions, all these 22 directions are supposed to have equal weights. According to formulas (9.02) and (9.03), the adjustment comprises six angle conditions and two side conditions.

Table IIa gives list of the observed triangles. The symbols Δ (number of the triangle), P (name of the station) and V (angle as the difference of two directions) refer to Fig. 10. Column "Observed" gives the observed values of the angles as computed from Table I. Symbol ε gives the spherical excess of the triangle, symbol w means the misclosure = sum of angles minus $(180^{\circ} + \varepsilon)$. For each angle, "log sin" has been found with 8 decimal places, at the same time, it's difference "diff" is written with 2 decimal places in units of the sixth decimal place per one second of arc.

The columns *g* and *h* show how the misclosures of the side conditions can be formed by formula (9.04). In a similar way, the columns F1 and F2 show how the logarithm of the ratio of both diagonals can be computed in two different ways by the aid of "log sin" of the observed angles.

Table I *Directions*

V	*P*	Observed
1	Mu–Ru	348° 43′ 56,″65
2	Ah	0
3	Km	36 17 49,84
4	Ah– Km	316 15 06,48
5	Mu	0
6	Ru	9 30 08,44
7	Ru– Ah	291 40 28,49
8	Hi	348 38 46,86
9	Km	0
10	Li	12 32 14,07
11	Mu	90 54 16,75
12	Km–Mu	318 28 09,60
13	Li	330 34 37,06
14	Ru	0
15	Hi	25 36 07,00
16	Ah	58 25 26,91
17	Li–Ru	284 55 09,40
18	Hi	0
19	Km	62 57 32,76
20	Hi–Km	297 59 02,73
21	Li	0
22	Ru	81 01 42,23

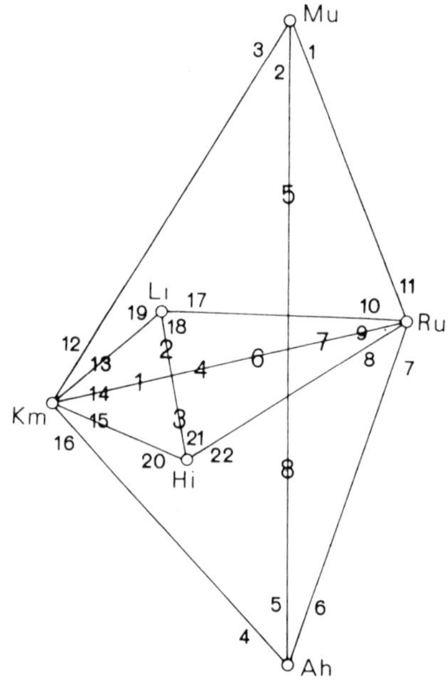

Fig. 10
Base extension net

Each quadrilateral gives three independent angle conditions. They are formed as a sum or a difference of two triangles in the following way:

Condition	Triangles	Misclosure
a	1 + 4 or 2 + 3	- 0".126
b	1 - 2 or 3 - 4	- 0.367
c	1 - 3 or 2 - 4	+ 0.355
d	5 + 8 or 6 + 7	- 0.1325
e	5 - 6 or 7 - 8	- 0.0365
f	5 - 7 or 6 - 8	+ 0.0365

The side conditions g and h are composed as explained in Chapter 9.3 by the aid of the list of triangles. Both the coefficients and the misclosures have been multiplied by 0.1 before they have been written in Table III. In the same way, the columns F1 and F2 are filled up, in order to compute the variance of the main side Ah - Mu. The column Σ of Table III is computed for the row sum checks. Note that in each column the sum of coefficients must be zero for each station.

In Table IV, the normal equations are given and solved by the fill matrix scheme. In order to make the sum check column $\Sigma 2$ more effective, the misclosures have been multiplied by 10.

As the result of the matrix computations, the following values are obtained. The square sum of the residuals is

$$- \mathbf{d}^T \mathbf{c} = - \mathbf{k}^T \mathbf{w} = 0.09085$$

The inverse weight of the logarithm of the fifth decimal place of the logarithm of rate Ah -.Mu/Li - Hi is obtained three times as the elements of the matrix $\mathbf{F}^T\mathbf{F} - \mathbf{D}^T\mathbf{C}$. The both values of the matrices $\mathbf{k}^T\mathbf{G} = - \mathbf{d}^T \mathbf{C}$ mean the corrections to the "observed" values $F1$ and $F2$ of this logarithm; hence, the adjusted value in 0.8065 6298. When the correlates $\mathbf{k}^T$ are inserted in the condition equations, the residuals $\mathbf{v}$ are obtained as shown in the last column of Table III where also $[\nu\nu] = 0.0907$ has been computed as a check. The variance of one direction is $m^2 = 0.0908/8 = 0.01135$ or $m = \pm 0''.1065$.

In Table IIb, column ν gives the corrections to the observed angles and column "ν.diff" the corrections to the values of "log sin". The next column gives the differences of the addidaments.

Table IIa *Triangles before the adjustment*

△	*P*	*V*	Observed	log sin	diff.	*g*	*h*	*F*1	*F*2
1	Km	15–13	55°01′29,″94	9.9134 9708	1,47			–	
	Li	19–18	62 57 32,76	9497 2278	1,08	–			
	Hi	21–20	62°00 57,27	9459 9902	1,12	+		+	
			ε 0,030						
			w –0,069						
2	Ru	10– 9	12 32 14,07	9.3366 0811	9,47	–		–	
	Km	14–13	29 25 22,94	6913 0613	3,73	+			
	Li	19–17	138 02 23,36	8251 7543	–2,34			+	
			ε 0,072						
			w +0,298						
3	Hi	22–20	143 02 39,50	9.7790 1700	-2.80				+
	Ru	9– 8	11 21 13,14	2941 6691	10,48	+			
	Km	15–14	25 36 07,00	6356 0070	4,40	–			–
			ε 0,064						
			w –0,424						
4	Li	18–17	75 04 50,50	9.9851 0727	0,56	+			+
	Hi	22–21	81 01 42,23	9946 5397	0,33	–			
	Ru	10– 8	23 53 27,21	6074 5100	4,76				–
			ε 0,097						
			w –0,057						
5	Mu	3– 1	47 33 53,19	9.8680 8028	1,93			–	
	Km	14–12	41 31 50,40	8215 2717	2,38		+	+	
	Ru	11– 9	90 54 16,75	9999 4587	–0,03		–		
			ε 0,406						
			w –0,066						
6	Ah	5– 4	43 44 53,52	9.8397 8613	2,20		+		
	Mu	3– 2	36 17 49,84	7723 0225	2,87		–		–
	Km	16–12	99 57 17,31	9934 1172	–0,37				+
			ε 0,700						
			w –0,030						
7	Ru	11– 7	159 13 48,26	9.5497 5865	–5,55			+	
	Ah	6– 5	9 30 08,44	2177 1540	12,58		–	–	
	Mu	2– 1	11 16 03,35	2909 0576	10,56		+		
			ε 0,153						
			w –0,103						
8	Km	16–14	58 25 26,91	9 9304 1290	1,29		–		
	Ru	9– 7	68 19 31,51	9681 5431	0,84		–		+
	Ah	6– 4	53 15 01,96	9037 7319	1,57				–
			ε 0,446						
			w –0,066						
			w_g	–623					
			w_h	–305					
			*F*1	0.8065 5940					
			*F*2	8065 6316					

Table IIb *Triangles after the adjustment*

V	*v*	*v.* diff.	3−δ	log **side**	**side**
15−13	0,″029	+ 4	0	3.6055 5309	Li–Hi **(base)**
19−18	−,042	− 5	0	3.6417 7870	Km–Hi
21−20	+,082	+ 9	0	3.6380 5508	Km–Li
	+,069				
10− 9	−,278	−263			
14−13	+,040	+ 15	− 14	3.9927 5574	Ru–Li
19−17	−,060	+ 14	− 29	4.1266 2488	Ru–Km
	−,298				
22−20	+,122	− 34			
9− 8	+,313	+328	0	3.6417 7870	Hi–Km
15−14	−,011	− 5	− 13	3.9832 0903	Hi–Ru
	+,424				
18−17	−,018	− 1			
22−21	+,040	+ 1	− 14	3.9927 5574	Li–Ru
10− 8	+,035	+ 17	0	3.6055 5307	Li–Hi
	+,057				
3− 1	+,112	+ 22	− 29	4.1266 2388	Km–Ru
14−12	+,085	+ 20	− 23	4.0800 7181	Mu–Ru
11− 9	−,131	0	− 55	4.2584 8999	Mu–Km
	+,066				
5− 4	+,029	+ 6			
3− 2	−,025	− 7	− 40	4.1910 0613	Ah–Km
16−12	+,026	− 1	−115	4.4121 1491	Ah–Mu **Main side**
	+,030				
11− 7	+,047	− 26			
6− 5	−,082	−103	− 23	4.0800 7181	Ru–Mu
2− 1	+,137	+145	− 33	4.1532 6455	Ru–Ah
	+,102				
16−14	−,059	− 8			
9− 7	+,178	+ 15	− 40	4.1910 0612	Km–Ah
6− 4	−,053	− 8	− 29	4.1266 2488	Km–Ru
	+,066				
	g	+623			
	h	+304			
	*F*1	+357	−115	4.4121 1491	**Main Side**
	*F*²	− 18	−115	4.4121 1492	25 829.436 m
				±49	±29

Table III *Condition equations*

v	*a*	*b*	*c*	*d*	*e*	*f*	*g*	*h*	F_1	F_2	*Σ*	*v*
1				−1	−1			−1,056	+0,193		−2,863	−0,083
2					+1	−1		+1.343		0,287	+1,630	+ ,054
3				+1		+1		− ,287	−,193	−,287	+1,233	+ ,029
4				−1	+1			− ,220		+,157	− ,063	+ ,008
5					−1	+1		+1,478	+1,258		+2,736	+ ,037
6				+1		−1		−1,258	−1.258	− ,157	−2,673	− ,045
7				−1		+1		− ,084	+ ,055	− ,084	+ ,387	+ ,002
8	−1		+1				−1,048			+ ,476	− ,572	− ,133
9		+1	−1		−1	−1	+1,995	+ ,081	+ ,947	+ ,084	+1,107	+ ,180
10	+1	−1					− ,047		− ,947	− ,476	−2,370	− ,098
11				+1	+1			+ ,003	− ,555		−1,448	+ ,049
12				−1		−1		−,238	− ,238	+ ,037	−2,439	− ,046
13	−1		−1				− ,373		+ ,147		−2,226	− ,001
14		−1	+1		+1	+1	+ ,813	+ ,367	+ ,238	+ ,440	+3,858	+ ,039
15	+1	+1					− ,440		− ,147	− ,440	+ ,973	+ ,028
16				+1	−1			− ,129		− ,037	− ,166	− ,020
17	−1	+1					− ,056		+ ,234	− ,056	+ ,122	+ ,026
18		−1	−1				+ ,164			+ ,056	−1,780	+ ,008
19	+1		+1				− ,108		− ,234		+1,658	− ,034
20	−1	−1					− ,112		− ,112	+ ,280	−1,944	−,068
21		+1	+1				+ ,145		+ ,112		+2,257	+ ,014
22	+1		−1				− ,033			− ,280	− ,313	+ ,054
								[*Fv*]	+ 358	− 18	[*vv*]	= 0,0907

Starting from the logarithm of the base line, definitive lengths of all other sides of the net have been computed in column "log side".

The standard error of the logarithm of side Ah - Mu is $0.1065 \cdot \sqrt{0.209} = 0.0487$ in units of the fifth decimal place, that of the side itself ± 29.0 mm. It has been supposed that the base line Li - Hi is errorless; therefore the standard error of side Ah - Mu is entirely due to the angle measurements of the extension net.

Exercise: The base extension net consists of n uniform rhombuses. In each rhombus, the ratio of the longer diagonal to the shorter is $K = \cot \alpha$. At each station, all directions have been observed in one set by the series method with equal weights.

Supposing that the standard mean error of one direction is m radians, compute the relative mean error of the longest diagonal D.

Answer: $$\frac{m_D}{D} = \sqrt{\frac{n}{2}}\,\frac{m}{\sin 2\alpha}$$

16.3 *Coordinate conditions*

When a chain of triangles forms a loop or when a chain of the second order must be adjusted between points of the first order triangulation, usually four "long" condition equations are brought forth, in addition to those "short" equations which are caused by individual triangles or by diagonal sides. One of these long equations expresses the fixed difference between the directions of the end sides; in this equation the coefficients are + or - 1. Another equation is a usual side equation, see (9.05). The remaining two equations express the coordinate conditions, or the fixed x- and y-coordinate differences. Here the coordinate conditions will be formed for the computations on a projection plane; the corresponding conditions on the ellipsoid are rather complicated and rarely needed.

First, approximate coordinates must be computed for each point of the new chain. This computation must be carried out with a final precision by the aid of a minimum number of observed angles. For this purpose, all diagonals and other extra sides are omitted, and in each of the triangles of the single chain two angles are selected. Starting from one fixed side (or two fixed neighboring points), the angles of each triangle can be denoted as follows:

forward angle α: opposite to the next connecting side (or to the end side),

backward angle γ: opposite to the previous connecting side (or to the initial side),

transverse angle β: opposite to the side along the right hand or left hand border of the chain.

All sides and coordinates are computed using α- and γ-angles only; for the checking computations, the observed values

Table IVa *Normal equations* (matrices **M, w, G**)

	a	*b*	*c*	*d*	*e*	*f*
a	8					
b		8			−2	−2
c			8		+2	+2
d				8		
e					8	
f						8
g						
h						
$\mathbf{F}^T\mathbf{F}$						
	End equations (matrices **R,T,C,** $-\mathbf{D}^T$, $\mathbf{k}^T$					
a	8					
b		8			−2	−2
c			8		+2	+2
d				8		
e		+.2500	−.2500		7	−1
f		+.2500	−.2500		+.1429	6.857
g	−.0076	−.2158	+.2430		+.0375	+.0438
h		+.0358	−.0358	+.0091	−.1394	−.2249
$-10\ \mathbf{d}^T$	+.1575	+.4588	−.4438	+.3656	+.3100	+.2552
$-\mathbf{D}^T$	+.1986	−.2464	+.1218	+.3145	+.2828	−.4172
	+.2370	+.0890	−.1320	+.0739	−0.564	+.0828
$10\ \mathbf{k}^T$	+.1520	+.4475	−.4120	+.1686	+.3227	+.2142

of β- angles are replaced by $180^\circ - \alpha - \gamma$ (or $\beta - w$, where w is the misclosure of the triangle). At the other end of the chain, there is a fixed point P which, in this way, reveals the coordinate misclosures w_x and w_y. Similarly, the end side reveals misclosures w_t in the direction and w_s in the length of the side.

The coefficients of the linear equations of coordinate conditions can be derived as follows. If the angles α and γ obtain small additions $d\alpha$ and $d\gamma$, the former causes an addition and the latter a diminution to the next connecting side and, in the same proportion, to the entire part of the chain from the point of angle β forward. The correction is proportional to

1. variations $d\alpha$ or $d\gamma$ themselves,
2. functions $\cot\alpha$ or $-\cot\gamma$,

Table IVb

g	*h*	10 **w**	**G**		*Σ*1	*Σ*2	
+0.061		−1.260	−1.597	−1.896	+4.568	+3.308	0
+1.736	−0.286	−3.670	+1.967	−0.712	+6.695	+3.025	0
−1.951	+0.286	+3.550	−0.978	+1.056	10.413	13.963	0
	−0.073	−1.325	−2.516	−0.591	4.820	3.495	0
−1.182	+1.119	−0.365	−2.715	+0.837	6.059	5.694	0
−1.182	+1.546	+0.365	+2.407	−0.182	10.589	10.954	0
7.045	+0.460	−6.230	+3.030	+0.661	8.669	2.438	−1
	7.038	−3.050	+3.466	+0.802	14.358	11.308	0
			5.940	1.234	2.062		−1

g	*h*	10 **c**	**C**		*Σ*3	
+0.061		−1.260	−1.597	−1.896	3.308	
+1.726	−0.286	−3.670	+1.967	−0.712	3.025	
−1.951	+0.286	+3.550	−0.978	+1.056	13.963	
	−0.073	−1.325	−2.516	−0.591	3.495	
−0.263	+0.976	−2.170	−1.979	+0.395	2.959	0
−0.300	+1.543	−1.750	+2.860	−0.568	8.642	−1
6.173	+0.696	−4.721	+2.430	+1.076	5.654	0
− .1127	6.456	−2.093	+2.908	+0.685	7.955	+1
+ .7647	+ .3242	9.085	+3.586	−0.169		
− .3937	− .4504		0.209	0.209		
− .1744	− .1061			0.209		
+ .7282	+ .3242		+3.57	−0.18		

3. coordinate differences $x_P - x_\beta$ and $y_P - y_\beta$, respectively.

The factor of proportionality is 1, if $d\alpha$ and $d\gamma$ are taken in radians; if the latter are in seconds of arc, the correction of the coordinate must be multiplied by ρ''.

On the other hand, $d\alpha$ and $d\gamma$ rotate the next side and the entire end part of the chain around the point of angle β. Rotation goes clockwise if angle β lies at the left hand border of the chain. Therefore, we mark such triangles by sign +, those triangles in which angle β lies at the right hand border, by sign −. The rotation causes corrections to the coordinates of P in such a way that dx is proportional to $y_\beta - y_P$ and dy proportional to

$x_P - x_\beta$.

Summing up all corrections, the coordinate conditions can be written in form:

$$\Sigma\left\{d\alpha\left[(x_P-x_\beta)\cot\alpha\mp(y_P-y_\beta)\right] + d\gamma\left[-(x_P-x_\beta)\cot\gamma\mp(y_P-y_\beta)\right]\right\} + \rho''\omega_x = 0$$
$$\Sigma\left\{d\alpha\left[(y_P-y_\beta)\cot\alpha\pm(x_P-x_\beta)\right] + d\gamma\left[-(y_P-y_\beta)\cot\gamma\pm(x_P-x_\beta)\right]\right\} + \rho''\omega_y = 0 \qquad (16.01)$$

The summation must be extended over all triangles of the chain. For triangles with sign +, the upper signs of formula (16.01) must be used, for triangles with sign −, the lower signs. If the coordinate differences are taken in units of 100 km and the misclosures in meters, the numerical values of ρ'' are 2.06265 sexagesimal or 6.36620 centesimal seconds.

The numerical computations can be arranged into schemes as shown by the following tables. As explanations to these tables, some remarks are necessary. In Table I, the *observed* values have already been reduced on to the projection plane; the two last columns of the table are filled up after the adjustment. In Table IIa, the angles of each triangle are in order α, β, γ. The numerical value of Sine has been sought for angles α and γ only, on the fourth row, the value of the connecting side is shown. In column Cot, the sign of $\cot\gamma$ has been altered as required by formula (16.01). The columns Δx and Δy contain the coordinate differences of formula (16.01) on the row of angle β; these differences have been multiplied by $\cot\alpha$ on the previous row and by $-\cot\gamma$ on the next row. After that, the sign of the original Δy has been changed in triangles with sign +, that of the Δx in triangles with sign −. The coefficients in column X and Y are obtained adding to the figures of the two previous columns on the same row those of the middle row but interchanging the latter. Table IIb gives a scheme for the computation of coordinates. The formulas of intersection have been used, alternatively either

$$x_\gamma - x_\beta = \frac{y_\beta - y_\alpha - (x_\beta - x_\alpha)\tan t_{\alpha\gamma}}{\tan t_{\alpha\gamma} - \tan t_{\beta\gamma}}$$
$$y_\gamma - y_\beta = (x_\gamma - x_\beta)\tan t_{\beta\gamma} \qquad (16.02a)$$

or

$$y_\gamma - y_\beta = \frac{x_\beta - x_\alpha - (y_\beta - y_\alpha)\cot t_{\alpha\gamma}}{\cot t_{\alpha\gamma} - \cot t_{\beta\gamma}} \qquad (16.02b)$$

$$x_\gamma - x_\beta = (y_\gamma - y_\beta)\cot t_{\beta\gamma}$$

The differences of the coordinates are written on the row of each α-angle; as a check, the next connecting side could be computed as the square root of their square sum.

The condition equations, Table III, can now be easily formed by the aid of figures and symbols of Table IIa. Only the equation of the fixed direction would be rather long because it contains all angles α and γ, or all directions of the chain. Therefore, it has been formed and checked in the following way along the right hand border of the chain:

Po - Sa	179°15'39".7		
2 - 1	55 45 34.5	(w)	+1".9
10 - 6	226 58 07.8	w_2	-3.3
21 - 18	187 11 21.1	w_3	-2.9
28 - 26	70 54 38.7	w_5	+4.6
01 - Ki	180 05 21.8	w_7	+2.8
w_t	+ 3.1		+3.1

The column v of Table III has been computed after the adjustment.

In the course of the adjustment, the weight coefficients and the error ellipse have been computed for the coordinates of point Ni at the middle of the chain. For this purpose, Tables II add. and III add. give columns to Tables II and III, respectively. The computation has been carried out by formulas (16.01). In the upper part of Table II add. the summation is started from the initial side, in the lower part, from the end side, and extended , in both cases, to point P = Ni.

The normal equations, Table IV, are formed in the usual way and solved by the fill matrix scheme. The additional columns give the weight coefficients for point Ni:

$Q_{xx} = 0.0871 \quad Q_{xy} = -0.0028 \quad Q_{yy} = 0.1939$

Using formulas (14.07) the error ellipse has been computed:

$w \sin 2t =$	-0.0056	$\tan 2t =$	0.524
$w \cos 2t =$	-0.1068	$2t =$	$183^{g}.0$
$w =$	0.1069	$t =$	$91^{g}.5$
$Q_{max} =$	0.1940	$m_{max} =$	$0''.616 = 0.299$ meters
$Q_{min} =$	0.0870	$m_{min} =$	$0.419 = 0.203$ "

Table V is the recomputation of Table IIb after the adjustment.

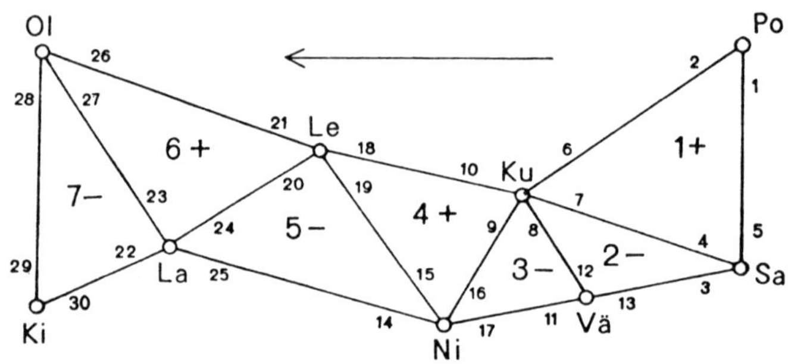

Fig. 11 *Second order triangulation in Kittilä*

Fixed sides:

Pomovaara x =	7540 084.70	y = 600 913.84
Sattasvaara	7504 588.30	601 371.68
s =	35 499.35	t = 179°15'39".7
Olostunturi x =	7537 584.97	y = 492 162.20
Kiuvaskero	7496 502.99	492 098.73
s =	41 082.03	t = 180°05'18".7

Table I *List of directions*

V	P → P	Observed	Adjusted	t
1	Po Sa	0°00′00,″0	00,0	179°15′39,″7
2	Ku	55 45 34,5	34,7	235 01 14,4
3	Sa Vä	0	0	258 31 32,0
4	Ku	29 17 21,7	24,5	287 48 56,5
5	Po	100 44 05,5	07,7	359 15 39,7
6	Ku Po	0	0	55 01 14,4
7	Sa	52 47 44,0	42,1	107 48 56,5
8	Vä	91 14 35,3	33,7	146 15 48,1
9	Ni	152 03 57,8	57,4	207 05 11,8
10	Le	226 58 07,8	05,9	281 59 20,3
11	Vä Ni	0	0	255 47 48,6
12	Ku	70 27 58,1	59,5	326 15 48,1
13	Sa	182 43 41,8	43,4	78 31 32,0
14	Ni La	0	0	285 45 12,1
15	Le	40 24 42,8	43,0	326 09 55,1
16	Ku	101 20 00,0	*59,8	27 05 11,9
17	Vä	150 02 36,5	36,5	75 47 48,6
18	Le Ku	0	0	101 59 20,4
19	Ni	44 10 36,4	34,7	146 09 55,0
20	La	133 45 10,6	07,4	235 44 27,8
21	01	187 11 21,1	20,2	289 10 40,6
22	La Ki	0	0	244 57 40,1
23	01	81 58 52,8	51,7	326 56 31,7
24	Le	170 46 46,2	47,7	55 44 27,8
25	Ni	220 47 33,8	32,0	105 45 12,1
26	01 Le	0	0	109 10 40,6
27	La	37 45 51,5	51,1	146 56 31,7
28	Ki	70 54 38,7	38,1	180 05 18,7
29	Ki 01	0	0	0 05 18,7
30	La	64 52 22,8	21,4	64 57 40,1

Table IIa *List of triangles*

	P	V	Observed	Sin side	Cot	△x	△y	X	Y
				35499,35					
1	Po	2– 1	55°45′34″,5	,826 6839.	+0,681	+0,225	−0,743	+1,317	−0,413
+	Sa	5– 4	71 26 43,8			+0,330	+1,092		
	Ku	7– 6	52 47 44,0	,796 4830	−0,759	−0,250	+0,829	+0,842	+1,159
			w = +2,3	36845,41					
2	Sa	4– 3	29 17 21,7	,489 2205	+1,783	+0,387	−1,322	−0,354	−1,539
–	Ku	8– 7	38 26 51,3			−0,217	−0,741		
	Vä	13–12	112 15 43,7	,925 4603	+0,409	+0,089	−0,304	−0,652	−0 521
			w = −3,3	19477,37					
3	Vä	12–11	70 27 58,1	,942 4441	+0,355	+0,077	−0,263	−0,664	−0,480
–	Ku	9– 8	60 49 22,5			−0,217	−0,741		
	Ni	17–16	48 42 36,5	,751 3809	−0,878	−0,191	+0,651	−0,932	+0,434
			w = −2,9	24430,12					
4	Ku	10– 9	74 54 10,0	,965 4853	+0,270	+0,117	−0,170	+0,747	+0,265
+	Ni	16–15	60 55 17,2			+0,435	+0,630		
	Le	19–18	44 10 36,4	,696 8745	−1,029	−0,448	+0,648	+0,182	+1,083
			w = +3,6	33846,73					
5	Ni	15–14	40 24 42,8	,648 2779	+1,175	+0,180	−0,519	−0,261	−0,672
–	Le	20–19	89 34 34,2			−0,154	−0,442		
	La	25–24	50 00 47,6	,766 1928	−0,839	−0,129	+0,371	−0,570	+0,217
			w = +4,6	28637,82					
6	Le	21–20	53 26 10,5	,803 1945	+0,742	+0,233	−0,152	+0,438	+0,163
+	La	24–23	88 47 53,4			+0,315	+0,205		
	01	27–26	37 45 51,5	,612 4147	−1,291	−0,406	+0,264	−0,201	+0,579
			w = −4,6	37559,09					
7	La	23–22	81 58 52,8	,990 2227	+0,141				
–	01	28–27	33 08 47,2						
	Ki	30–29	64 52 22,8	,905 3688	−0,469				
			w = +2,8	41079,24					
			pro	41082,03					
			w_s	≐ − 2,79					
				= − 14″,0					

Table IIb *Computation of coordinates*

			− 35 496,40	+ 457,84
P	*t*	cot or tan	*x*	*y*
Sa	179°15′39,″7		7504 588,30	601 371,68
Po	235 01 14,2	+0,699 6716	+11 273,37	−35 078,42
Sa				
Ku	287 48 58,2	−0,321 3761	7515 861,67	566 293,26
		+1,021 0477		
Sa	258 31 36,5	+0,202 9651	−16 197,61	+10 816,90
Ku				
Vä	146 15 52,8	−1,497 4354	7499 664,06	577 110,16
		+1,700 4005		
Vä	255 47 54,7	+0,253 0662	−21 750,26	−11 124,61
Ku				
Ni	207 05 18,2	+1,955 1483	7494 111,40	555 168,65
		−1,702 0820		
Ku	281 59 28,2	−0,212 3954	+28 115,57	−18 844,52
Ni				
Le	326 10 04,6	−1,491 9758	7522 226,97	536 324,13
		+1,279 5804		
Ni	285 45 21,8	−0,282 1433	−16 120,47	−23 669,71
Le				
La	235 44 34,2	+0,681 0588	7506 106,51	512 654,42
		−0,963 2021		
Le	289 10 44,7	−0,347 8276	+31 479,48	−20 487,26
La				
01	326 56 36,2	−1,536 5391	7537 585,99	492 167,15
		+1,188 7115	w_x = +2,10	w_y = +10,21
La	244 57 43,4	+2,140 8043	−41 079,20	− 63,85
01				
Ki	180 05 20,6	+0,001 5543		
	(w = +1,9)	+2,139 2500		

Table III *Condition equations*

V	1	2	3	4	5	6	7	t	s	x	y	v
1	−1							−1	−0,681	−1,317	+0,413	−0,11
2	+1							+1	+0,681	+1,317	−0,413	+0,11
3		−1							−1,783	+0,354	+1,539	−1,68
4	−1	+1							+1,783	−0,354	−1,539	+1,15
5	+1											+0,53
6	−1							−1	+0,759	−0,842	−1,159	+1,15
7	+1	−1							−0,759	+0,842	+1,159	−0,76
8		+1	−1									−0,46
9			+1	−1					−0,270	−0,747	−0,265	+0,79
10				+1				+1	+0,270	+0,747	+0,265	−0,72
11			−1						−0,355	+0,664	+0,480	−1,00
12		−1	+1						−0,054	−0,012	+0,041	+0,42
13		+1							+0,409	−0,652	−0,521	+0,58
14					−1				−1,175	+0,261	+0,672	+0,01
15				−1	+1				+1,175	−0,261	−0,672	+0,20
16			−1	+1					+0,878	+0,932	−0,434	−0,23
17			+1						−0,878	−0,932	+0,434	+0,02
18				−1				−1	+1,029	−0,182	−1,083	+1,42
19				+1	−1				−1,029	+0,182	+1,083	−0,25
20					+1	−1			−0,742	−0,438	−0,163	−1,74
21						+1		+1	+0,742	+0,438	+0,163	+0,57
22							−1		−0,141			+0,34
23						−1	+1		+0,141			−0,78
24					−1	+1			+0,839	+0,570	−0,217	+1,87
25					+1				−0,839	−0,570	+0,217	−1,43
26						−1		−1	+1,291	+0,201	−0,579	+0,32
27						+1	−1		−1,291	−0,201	+0,579	−0,04
28							+1	+1				−0,28
29							−1		+0,469			+0,72
30							+1		−0,469			−0,72

$[vv] =$ 22,19
$m =$ ± 1",42

Tables II add. and III add.

Computation of the weight coefficients for point Ni

P	Δ*x*	Δ*y*	*X*	*Y*
Po	−0,071	−0,314	+0,391	−0,419
Sa	−0,105	+0,462		
Ku	+0,080	+0,351	+0,542	+0,246
Sa	−0,388	−0,198	−0,499	+0,019
Ku	+0,217	−0,111		
Vä	−0,089	−0,045	−0,200	+0,172
Vä	−0,077	−0,039	−0,188	+0,178
Ku	+0,217	−0,111		
Ni	+0,191	+0,098	+0,080	+0,315
Ku				
Ni				
Le				
Ni	+0,330	−0,221	+0,142	−0,502
Le	−0,281	−0,188		
La	−0,236	+0,158	−0,424	−0,123
Le	+0,089	−0,315	+0,514	−0,195
La	+0,120	+0,425		
01	−0,155	+0,549	+0,270	+0,669
La	+0,061	−0,089	−0,569	−0,524
01	−0,435	−0,630		
Ki	−0,204	+0,296	−0,834	−0,139

V	*X*	*Y*	*X*	*Y*
1	−0,391	+0,419		
2	+0,391	−0,419		
3	+0,499	−0,019		
4	−0,499	+0,019		
5				
6	−0,542	−0,246		
7	+0,542	+0,246		
8				
9				
10				
11	+0,188	−0,178		
12	+0,012	+0,006		
13	−0,200	+0,172		
14			−0,142	+0,502
15			+0,142	−0,502
16	−0,080	−0,315		
17	+0,080	+0,315		
18				
19				
20			−0,514	+0,195
21			+0,514	−0,195
22			+0,569	+0,524
23			−0,569	−0,524
24			+0,424	+0,123
25			−0,424	−0,123
26			−0,270	−0,669
27			+0,270	+0,669
28				
29			+0,834	+0,139
30			−0,834	−0,139

Table IV *Normal equations and the solution*

	1	2	3	4	5	6	7	*t*	*s*
1	6	−2						+3	−1,940
2		6	−2						+4,789
3			6	−2					−1,726
4				6	−2			+2	−1,815
5					6	−2			+0,959
6						6	−2	+2	−0,401
7							6	+1	+0,635
t								8	−0,707
s									21,617
x									
y									
X									
Y									
X									
Y									
1	6	−2						+3	−1,940
2	+,3333	5,333	−2					+1	+4,142
3		+,3750	5,25	−2				+0,375	−0,173
4			+,3810	5,238	−2			+2,143	−1,881
5				+,3818	5,236	−2		+0,818	+0,241
6					+,3819	5,236	−2	+2,312	−0,309
7						+,3820	5,236	+1,883	+0,517
t	−,5000	−,1875	−,0714	−,4091	−,1562	−,4416	−,3597	3,582	+0,181
s	+,3233	−,7766	+,0329	+,3591	−,0460	+,0589	−,0987	−,0506	17,003
x	−,7784	+,1188	+,6714	+,3266	+,3115	−,0804	−,0692	−,5348	−,2585
y	−,5051	+,7102	+,2396	−,4689	+,2328	−,1531	+,0521	−,1780	+,5517
w	−0,383	+0,475	+0,733	−0,407	−1,034	+0,484	−0,350	−0,402	+,665
k	+0,535	−0,530	−0,071	−0,211	−1,318	+0,445	−0,427	+0,145	+0,618
X	−,3939	+,1807	+,0721	+,0428	+,0163	+,0062	+,0024	−,1320	+,0565
Y	+,0610	+,0306	−,1435	+,0054	+,0021	+,0008	+,0003	+,1164	+,0736
X				+,0271	+,2164	−,4066	+,4321	−,2492	−,1198
Y				−,0959	+,1650	−,2413	+,2888	−,1085	+,1519

Table IV add.

	x	*y*	*w*	Σ	*X*	*Y*	*X*	*Y*
1	+4,671	+3,030	+2,3	15,061	+2,364	−0,366		
2	−2,190	−4,798	−3,3	−3,500	−1,752	−0,041		
3	−3.287	+0,163	−2,9	−5,751	−0,017	+0,814		
4	+3,053	+2,936	+3,6	11,774	−0,080	−0,315	−0,142	+0,502
5	−2,284	−2,157	+4,6	3,118			−1,079	−1,056
6	+1,044	+1,267	−4,6	1,311			+2,562	+1,594
7	+0,202	−0,579	+2,8	8,057			−3,075	−1,995
t	+4,642	+2,423	+3,1	25,458	+1,323	−0,592	+0,784	+0,474
s	+1,928	−14,240	−14,0	−4,901	−2,359	−1,296	+1,732	−3,007
x	10,176	+2,609	+2,10	22,661	+2,401	−1,522	+0,752	−0,038
y		12,854	10,21	13,715	+2,733	+0,956	+0 106	+1,334
X					1,478	−0,098		
Y						0,733		
X							3,112	+0,950
Y								2,094
1	+4,671	+3,030	+2,300	15,061	+2,364	−0,366		
2	−0,634	−3,788	−2,533	+1,521	−0,964	−0,163		
3	−3,525	−1,258	−3,850	−5,180	−0,378	+0,753		
4	+1,711	+2,456	+2,133	9,800	−0,224	−0,028	−0,142	+0,502
5	−1,631	−1,219	+5,414	6,860	−0,086	−0,011	−1,133	−0,864
6	+0,421	+0.802	−2,532	+3,931	−0,033	−0,004	+2,129	+1,264
7	+0,362	−0,273	+1,833	+9,558	−0,012	−0,002	−2,262	−1,512
t	+1,916	+0,638	+1,440	+7,758	+0,473	−0,417	+0,893	+0,389
s	+4,395	−9,380	−11,302	+0,716	−0,960	−1,251	+2,036	−2,583
x	0,811	−0,190	+0,642	+1,264	+0,238	−0,198	−0,574	−0,008
y	+,2336	1,426	+0,729	+2,155	+0,295	+0,554	+0,295	−0,872
w	−0,791	−0,511	22,213					
k	−0,911	−0,511	22,213					
X	−,2929	−,2068			0,087	−0,003		
Y	+,2438	−,3887				0,193		
X	+,7069	−,2070					0,087	−0,003
Y	+,0101	+,6111						0,195

Table V *Compution of the adjusted coordinates*

P	t	cot	x	y
Sa			7504 588,30	601 371,68
Po	235°01′14,″4	+0,699 6700	+11 273,17	−35 078,78
Sa				
Ku	287 48 56,5	−0,321 3671 +1,021 0371	7515 861,47	566 292,90
Sa	258 31 32,0	+0,202 9879	−16 197,89	+10 817,62
Ku				
Vä	146 15 48,1	−1,497 3617 +1,700 3496	7499 663,58	577 110,52
Vä	255 47 48,6	+0,253 0977	−21 751,35	−11 124,32
Ku				
Ni	207 05 11,9	+1,955 2965 −1,702 1988	7494 110,12	555 168,58
Ku	281 59 20,4	−0,212 3557	+28 115,84	−18 846,58
Ni				
Le	326 09 55,1	−1,491 8269 +1,279 4712	7522 225,96	536 322,00
Ni	285 45 12,1	−0,282 0924	−16 122,09	−23 670,51
Le				
La	235 44 27,8	+0,681 1043 −0,963 1968	7506 103,87	512 651,49
Le	289 10 40,6	−0,347 8052	+31 481,09	−20 489,29
La				
01	326 56 31,7	−1,536 4662 +1,188 6610	7537 584,96	4 92 162,20
La	244 57 40,1			
01				
Ki	180 05,18,7			

16.4 *Adjustment of linear functions*

An empirical function can often be idealized by equation $y = s + kx$, where the free variable x is supposed to have a series of exactly known values and y a corresponding series of observed values; the latter may be characterized by mean errors or by weights. The unknown parameters, sum constant s and scale factor k are determined by an adjustment. There are many practical applications in which often only the factor k is important, *e.g.* calibration of measuring instruments, determination of thermal coefficients or the rate of a clock, velocity of movements or phenomena, regression analysis.

Supposing that there are n equally weighted observations, the observation equations $v = s + kx - y$ give two normal equations

$$n\,s + [x]\,k - [y] = 0$$
$$[x]\,s + [x^2]\,k - [xy] = 0$$

The solution can be obtained readily using determinant D:

$$D = n\,[x^2] - [x]^2 \qquad (16.03)$$
$$s = ([x^2][y] - [x]\,[xy])/D$$
$$k = (n[xy] - [x][y])/D$$

The weight coefficients of s and k are

$$Q_{ss} = [x^2]/D$$
$$Q_{kk} = n/D \qquad (16.04)$$
$$Q_{sk} = -[x]/D$$

and the square sum of the residuals

$$[vv] = [y^2] - s[y] - k[xy] \qquad (16.05)$$

Standard error of one observation is $m = \sqrt{[vv]/(n-2)}$.

Denoting the arithmetical mean values of x and y by x_m and y_m, resp., the parameters can be computed by formulas

$$k = [(x - x_m)(y - y_m)]/[(x - x_m)^2] \qquad (16.06)$$
$$s = y_m - kx_m \qquad (16.07)$$

The adjustment of a linear function can be made easier if the values of the free variable x can be chosen equally spaced. If the number n of the observations is odd, use for x following values

$$0,\ \pm 1,\ \pm 2, \ldots \frac{n-1}{2}$$

if n is even, following values

$$\pm \frac{1}{2},\ \pm \frac{3}{2}, \ldots \pm \frac{n-1}{2}$$

thus taking one interval as the unit. In both cases, the square sum of these values is

$$N = \frac{(n - 1)n(n + 1)}{12} \tag{16.08}$$

Scale factor k can now be computed in the following way. After the column of y, a new column is written for differences: last y − first y, last but one y − second y, third y from the end − third y a.s.o. If n is odd, the central value of y is not used. Starting from the smallest differences, they are multiplied by 1, 2, ... $\frac{n-1}{2}$ if n is odd and by 1, 3, 5 ...$n - 1$ if n is even. The sum of the former products is divided by N, that of the latter by $2N$. According to (16.06), the quotient is $= k$ for one interval. Number N is the weight of k.

Example 1: *Calibration of a levelling scale:*

x	y	Δy	Product	v
0"	- 3.7			-.02
10	- 2.3			-.07
20	- 1.0			-.03
30	+ 0.3			+.02
40	1.6			+.06
50	3.0			+.01
60	4.3	2.7	2.7	+.05
70	5.6	5.3	10.6	+.10
80	7.0	8.0	24.0	+.05
90	8.5	10.8	43.2	-.11
100	9.8	13.5	67.5	-.06
Sum	33.1		148.0	0

Here $n = 11$, $N = 110$, $k = 148.0/110 = 1.3455$ parts/10". One part is $10''/k = 7''.432$. From $[vv] = 0.0418$ can be computed: $m^2 = 0.0418/9 = 0.00465$, $m_{kk} = m^2/110 = 0.000042$, $m_k = \pm\ 0.0065$ and the standard error of the value of one part $10'' m_k/k^2 = \pm\ 0''.036$. Note, however, that the individual values of v shov a systematical wavy character; therefore, the standard error can not be considered reliable.

Example 2: *Linear interpolation by adjustment.*

This example shows, how from a table with 7 decimal places a correc value with 8 decimal places could be found. We take a fragment from logarithmic tables:

x	y = log x	diff.
27178	434 2175	
79	2335	160
80	2495	160
81	2654	159
82	2814	160
83	2974	160
84	3134	160
85	3293	159
86	3453	159
87	3613	160
88	3773	160
Sum	32713	1598
Mean	434 297391	159.8

The value of μ = log 2.7182 82828 is interpolated supposing, that the tabulated values are uncorrelated equally weighted "observations". Here k is a less important parameter, therefore we can use the approximate value k = 159.8. The "adjusted" value of y for x = 27183 is the mean value 434 2973.91. Subtracting $k \cdot dx$= 0.17172k = 29.04, the interpolated value 434 2944.87 is obtained. In fact, the correct value is 434 2944.82. Note, however, that systematic rounding-off errors in the table may render the extra decimal place quite illusory.

As mentioned in Chapter 11.5 the formulas of this chapter give the direction of the regression line of y with respect to x. If the values of y are regarded errorless and the residuals v are added to the values of x, the regression line of x with respect to y is obtained. In some applications, both x and y must be considered erroneous or stochastic variables. By a change of scale, they can be made equally weighted. In the graphical presentation, we draw a straight line through the group of points in such a way that the perpendicular distances v of all points from this line comply to the principle of least squares. It can be shown that the line goes through the center of gravity (with x_m and y_m as coordinates) which now is taken as origin. The direction t of the line can be obtained by formulas

$$\begin{cases} D \sin 2t = 2[xy] \\ D \cos 2t = [x^2] - [y^2] \\ v = x \sin t - y \cos t \\ [vv] = \frac{1}{2}\left\{[x^2] + [y^2] - D\right\} \\ p_t = \frac{1}{2}\left\{[x^2] + [y^2] + D\right\} \\ m_t = \sqrt{\dfrac{[vv]}{p_t(n-2)}} \end{cases} \qquad (16.09)$$

where n is the number of points.

The methods of this chapter can very easily be expanded for non-linear cases where through the points x, y a curve of higher degree is sought, *e.g.* a cubic parabola. In the observation equations

$$k_0 + k_1 x + k_2 x^2 + k_3 x^3 - y = v$$

the constants k are the parameters which must be determined.

Example 3:

The nucleus of an atom contains Z charged particles (protons) and N neutral particles (neutrons). The following table shows some chemical elements which have only one isotope. Compute the coefficients C_2 and C_4 for the approximate formula $N - Z = C_2 Z^2 + C_4 Z^4$.

Element	F	Na	P	Sc	Mn	As	Y	J	Tb	Cp	Au	Bi
$A = N + Z$	19	23	31	45	55	75	89	127	159	175	197	209
Z	9	11	15	21	25	33	39	53	65	71	79	83

Answer: $C_2 = 79.6 \pm 1.9$ $C_4 = 26.1 \pm 3.2$ $m = \pm 0.5$

16.5 *Adjustment of periodic functions*

Many empirical functions can be expressed by Fourier series

$$y = y_m + k_1 \sin(u - u_1) + k_2 \sin 2(u - u_2) + k_3 \sin 3(u - u_3) + \ldots \quad (16.10)$$

where u is the free variable, either an angle or just regarded as an angle the entire range of which is 360°. The values of u can usually be chosen with equal intervals: 0, $360°/n$, $2 \cdot 360°/n$, $(n - 1) \cdot 360°/n$. It is not necessary to decide beforehand how many terms or "waves" should be computed but the number of parameters y_m, k_i and u_i can not be greater than n.

The computation method is known as *harmonic analysis* and it is very much used in various sciences. In geodesy, it is used especially for the investigation of graduation circles.

For an adjustment, new parameters are inserted:

$$\begin{aligned} a_i &= -k_i \sin iu \\ b_i &= +k_i \cos iu \end{aligned} \quad (16.11)$$

Then the observation equations can be written in linear form

$$\begin{aligned} v = y_m &+ a_1 \cos u + a_2 \cos 2u + a_3 \cos 3u + \ldots \\ &+ b_1 \sin u + b_2 \sin 2u + b_3 \sin 3u + \ldots - y \end{aligned} \quad (16.12)$$

When the values of u cover the entire range from 0° to 360° with equal intervals, $[\cos iu] = [\sin iu] = 0$. It can also be shown that

$$\begin{aligned} [\cos(i \pm j)u] &= [\cos iu \cos ju] \mp [\sin iu \sin ju] = 0 \\ [\sin(i \pm j)u] &= [\sin iu \cos ju] \pm [\cos iu \sin ju] = 0 \end{aligned} \quad (16.13)$$

when $i \neq j$ and $i \pm j \neq n$. Therefore, the normal equations will be very simple because only the diagonal coefficients $[\cos^2 iu]$ and $[\sin^2 iu]$ remain and the other coefficients, being product sumns, disappear. Furthermore, because

$$\begin{aligned} [\cos 2iu] &= [\cos^2 iu] - [\sin^2 iu] = 0 \quad \text{when } i < \tfrac{n}{2} \\ [\cos(i - i)u] &= [\cos^2 iu] + [\sin^2 iu] = n \end{aligned} \quad (16.14)$$

we can conclude that all diagonal coefficients are $= \frac{n}{2}$. The entire system of normal equations can be written in form

$$\begin{aligned} n\, y_m &= [y] \\ \frac{n}{2} a_i &= [y \cos iu] \\ \frac{n}{2} b_i &= [y \sin iu] \end{aligned} \quad (16.15)$$

which gives an easy solution. The weight of the mean value y_m is n and the weight of each parameter a_i or $b_i = \frac{n}{2}$. The weight of each of the original coefficients k_i is also $= \frac{n}{2}$ and that of the parameters u_i is $= \frac{n}{2}(ik_i/\rho)^2$.

It is advisable to compute the individual terms $i = 1, 2, 3, \ldots$ one by one and, each time, write down the residuals y. For the computation of the next term $i + 1$, the original values of y can then be replaced by these residuals.

As an example, we take the time equation (difference of the true solar time and the mean time). As the observed values y, take the mean time at the moment when the sun dial shows noon at the beginning of each month. The computations are carried out in the following order:

$$\begin{aligned} d(u) &= y(u) - y(180^\circ + u) \\ y_s &= d(u) + d(180^\circ - u) \\ y_c &= d(u) - d(180^\circ - u) \\ s &= y_s \sin u \\ c &= y_c \cos u \end{aligned} \qquad \begin{aligned} \sin 0^\circ &= \cos 90^\circ = 0 \\ \sin 30^\circ &= \cos 60^\circ = 0.5 \\ \sin 60^\circ &= \cos 30^\circ = 0.866 \\ \sin 90^\circ &= \cos 0 = 1 \end{aligned}$$

Residuals v_1 after the first wave $i = 1$ show that a second wave $i = 2$ is essential. It has been computed in the second table in the order

$$\begin{aligned} e(u) &= y(u) - y(90^\circ + u) + y(180^\circ + u) - y(270^\circ + u) \\ y_S &= e(u) + e(90^\circ - u) \\ y_C &= e(u) - e(90^\circ - u) \\ S &= y_S \sin 2u \\ C &= y_C \cos 2u \end{aligned}$$

Residuals v_2 are hardly more than rounding off errors.

The standard error of one observation is $m = \sqrt{2.4/7} = \pm 0.6$. Dividing it by $\sqrt{6}$, the standard error of coefficients a, b and $k = \pm 0.24$ is obtained. Furthermore, standard error of u_1 is $= 1^\circ\!.9$ and that of u_2 is $= 0^\circ\!.7$.

According to spherical astronomy, u_1 = Jan. 4th means the time of perihelion and u_2 = Dec. 22th the time of winter solstice, k_1 = 0.032 = double eccentricity of the orbit of the earth and k_2 = 0.043 = $\tan^2\varepsilon/2$ (ε = obliquity of the ecliptic).

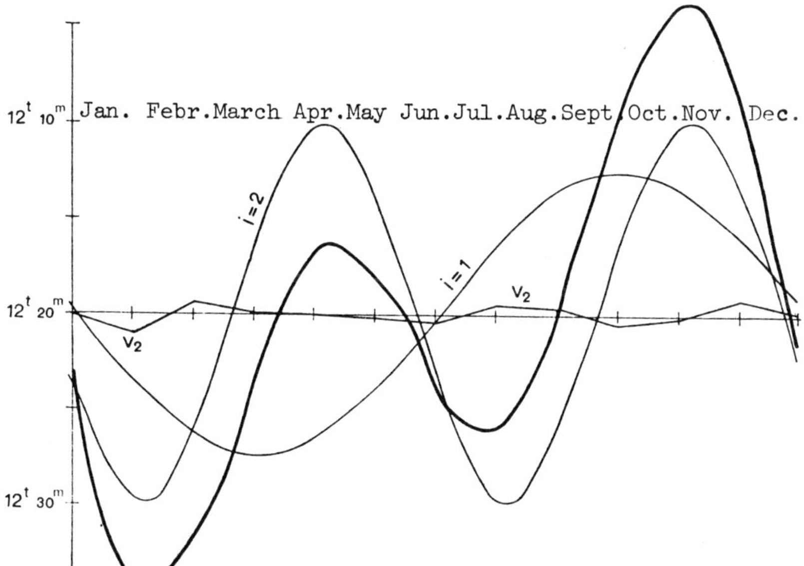

Fig. 12 *Time equation*

Time		u	y	d	y_s	y_c	s	c
Jan.	1	0°	12^t23^m	- 1		-1		-1
Febr.	1	30	12 34	+ 8	+ 17	-1	+ 8.5	-0.87
March	1	60	12 32	+12	+ 25	-1	+21.65	-0.5
April	1	90	12 24	+14	+ 14		+14	
May	2	120	12 17	+13				
June	1	150	12 18	+ 9	Sum		+44.15	-2.37
July	2	180	12 24					
Aug.	1	210	12 26				m	
Sept.	1	240	12 20			b_1	+7.358	
Oct.	1	270	12 10			a_1	-0.395	
Nov.	1	300	12 04			k_1	7.369	
Dec.	1	330	12 09			u_1	3°1	
y_m			12 20					

u	$-v_1$	e	y_s	y_c	S	C	$-v_2$
0°	+ 3.3	+13		+13		+13	0.0
30	+10.6	+39	+64	+14	+55.42	+ 7	+0.9
60	+ 5.7	+25					-0.6
90	- 3.4		Sum		+55.42	+20	-0.1
120	- 9.7			b_2	+ 9.237		0.0
150	- 6.1			a_2	+ 3.333		+0.2
180	+ 3.5			k_2	9.820		+0.2
210	+ 9.3			u_2	- 9°9		-0.4
240	+ 6.1						+0.2
270	- 2.7						+0.6
300	- 9.5						+0.2
330	- 7.1						-0.7
						[vv] =	2.4

16.6 *Coordinate transformations*

Suppose that there are two Cartesian coordinate systems for the same points of a plane. Let X, Y be the main system (general or definitive) and x, y the auxiliary system (local or temporary). If there are points for which the coordinates in both systems have been determined, various formulas for the coordinate transformation can be derived. In geodesy, the most important case is *Helmert transformation* or the linear conformal transformation. It consists of four parts: transfer of origin in x- and y-directions, rotation of the axes and change of scale. The transformation formulas read

$$\begin{aligned} X &= A + C\,x + D\,y \\ Y &= B + C\,y - D\,x \end{aligned} \tag{16.16}$$

Denoting

$$\begin{aligned} C &= K \cos T \\ D &= K \sin T \end{aligned} \tag{16.17}$$

the transformation can be interpreted geometrically as follows:

A, B = the new coordinates of the old origin

T = direction of the new X -axis in the old system

K = the scale (new measures/old measures).

These four parameters can be computed if the coordinates in both systems are given for two points. If the number n of points is greater, an adjustment is necessary. Suppose that the new coordinates X, Y are correct and that the old coordinates x, y need small random corrections or residuals v. In this case, the observation equations should be taken in form

$$\begin{aligned} a + c\,X - d\,Y - x &= v_x \\ b + c\,Y + d\,X - y &= v_y \end{aligned} \tag{16.18}$$

After the solution of parameters a, b, c, d, those of the definitive transformation formulas (16.16) can be computed by formulas

$$\begin{aligned} k^2 &= c^2 + d^2 \\ A &= (-ac-bd)/k^2 = -aC - bD \\ B &= (-bc+ad)/k^2 = -bC + aD \\ C &= c/k^2 \\ D &= d/k^2 \end{aligned} \tag{16.19}$$

In practice, however, the solution can be found easier inserting the residuals into the final transformation formulas (16.16):

$$\begin{aligned} A + Cx + Dy - X &= -Cv_x - Dv_y = v_X \\ B + Cy - Dx - Y &= Dv_x - Cv_y = v_Y \end{aligned} \qquad (16.20)$$

for the solution will the same, with a sufficient accuracy, when the square sum of the new residuals:

$$[v_X^2 + v_Y^2] = (C^2 + D^2)[v_x^2 + v_y^2]$$

is minimized, instead of the sum of squares of the true residuals.

From (16.20), which here are considered observation equations, with equal weights, the normal equations are formed. Denoting

$$\begin{aligned} p &= [x], \; P = [X], \quad N = [x^2 + y^2], \; R = [xX + yY] \\ q &= [y], \; Q = [Y], \quad M = nN - p^2 - q^2, \; S = [yX - xY] \end{aligned} \qquad (16.21)$$

the normal equations can be written in form

$$\begin{aligned} nA \qquad\quad + pC + qD &= P \\ nB + qC - pD &= Q \\ pA + qB + NC \qquad &= R \\ qA - pB \qquad + ND &= S \end{aligned} \qquad (16.22)$$

The solution is

$$\begin{aligned} A &= (NP \qquad -pR - qS)/M \\ B &= (\qquad NQ - qR + pS)/M \\ C &= (-pP - qQ + nR \qquad)/M \\ D &= (-qP + pQ \qquad + nS)/M \end{aligned} \qquad (16.23)$$

The matrix of the weight coefficients of the unknown parameters is

$$\mathbf{Q} = \begin{array}{c} \\ \end{array} \begin{bmatrix} N/M & & -p/M & -q/M \\ & N/M & -q/M & p/M \\ -p/M & -q/M & n/M & \\ -q/M & p/M & & n/M \end{bmatrix} \qquad (16.24)$$

(columns: A, B, C, D)

The original parameters

$$K = \sqrt{C^2 + D^2} \quad \text{and} \quad \tan T = D/C \qquad (16.25)$$

are uncorrelated; the standard errors are

$$m = \sqrt{\frac{[vv]}{2(n-2)}} \qquad m_K = m\sqrt{\frac{n}{M}} \qquad m_T = \frac{\rho}{K}\, m_K$$

The method described above gives the solution with the smallest amount of computational labor if a large computer with many digits is available. Working with smaller computers, sufficient accuracy can be maintained by the *method of gravity center.* For this method, compute first the arithmetical mean of each coordinate and take this "gravity center" as a new origin in both systems. From the formulas above, p, q, P and Q disappear and the solution is

$$C = R/N \quad \text{and} \quad D = S/N \tag{16.26}$$

Because the values of p, q, P and Q must be computed, anyway, for the original coordinates, the remaining parameters of the transformation formulas can now written in form

$$\begin{aligned} A &= (P - pC - qD)/n \\ B &= (Q - qC + pD)/n \end{aligned} \tag{16.27}$$

If the new coordinates X and Y differ only slightly from the old coordinates x and y, the parameters D and $C - 1 = E$ are small. In this case, write the small differences $\xi = X - x$ and $\eta = Y - y$. Replacing x and y by ξ and η in (16.21), formulas 16.23) give a correct solution except C will be replaced by E. The definitive transformation formulas read

$$\begin{aligned} X &= x + A + Ex + Dy \\ Y &= y + B + Ey - Dx \end{aligned} \tag{16.28}$$

As an example, consider the following table where x, y (in millimeters) have been measured from a rectified aerial photograph and X, Y (in meters) are the geodetic coordinates of the control points.

Point	x	y	X	Y	v_X	v_Y
1	591.82	233.07	10975.91	11690.49	+49	00
2	558.16	284.90	10581.18	12165.93	+34	+38
3	500.24	338.80	9943.05	12634.50	+12	-08
4	494.66	357.80	9866.16	12816.77	-58	-05
5	530.74	373.22	10206.25	13012.42	-11	-43
6	445.16	321.56	9415.69	12399.09	+24	-16
7	374.00	258.84	8781.55	11692.43	+35	+33
8	363.44	240.78	8698.20	11501.03	-19	-03
9	436.39	239.54	9424.36	11573.56	-01	+12
10	457.60	227.72	9649.55	11480.98	-66	-04
Sum	4752.21 $= p$	2876.23 $= q$	97541.90 $= P$	120967.20 $= Q$	-01	+04

$N = 3163\,572$ $\qquad A = 5367.077$

$R = 81921\,516$ $\qquad B = 8684.884$

$S = -29521\,691$ $\qquad C = +9.93697$

$M = 779\,521$ $\qquad D = -1.16522$

$K = 10.00505$

$T = -7^{g}.4311$

In the next table, the same problem has been computed again using the method of gravity center:

Point	x	y	X	Y	v_X	v_Y
1	+116.60	-54.55	+1221.72	-406.23	+49	00
2	+ 82.94	- 2.72	+ 826.99	+ 69.21	+35	+37
3	+ 25.02	+51.18	+ 188.86	+537.78	+12	-08
4	+ 19.44	+70.18	+ 111.97	+720.05	-58	-05
5	+ 55.52	+85.60	+ 452.06	+915.70	-11	-43
6	+ 30.06	+33.94	- 338.50	+302.37	+24	-17
7	-101.22	-28.78	- 972.64	-404.29	+35	+33
8	-111.78	-46.84	-1055.99	-595.69	-19	-04
9	- 38.83	-48.08	- 329.83	-523.16	-01	+11
10	- 17.62	-59.90	- 104.64	-615.74	-66	-05
Sum	+ 0.01	+ 0.03	0	0	0	-01

$[vv] = 1.86$

$N = 77\,952.0613$ $\qquad C = +\ 9.936\,9737$

$R = 774\,607.5838$ $\qquad D = -\ 1.165\,2134$

$S = -90\,830.7826$ $\qquad K^2 = 100.101\,169$

Computing $H = [x^2 + y^2] = 7803\,094.2890$, $[vv] = H - K^2N = 1.86$ is obtained as a check. The standard error of unit weight

$$m = \sqrt{1.86/16} = \pm\ 0.34 \text{ meters}$$

means here the accuracy of one coordinate as measured from the photograph but computed in the geodetic system.

In cartography and in photogrammetry, the conformal transformation is not as usual as the *affine transformation* in which the scale is different along x- and y- axes and in which the angle between the axes can be altered. The transformation formulas are

$$\begin{aligned} X &= A + Cx + Dy \\ Y &= B + C'y - D'x \end{aligned} \qquad (16.29)$$

where $C' \neq C$ and $D' \neq D$. Instead of one large system of normal equations, there are two smaller independent groups.

Helmert transformation and the affine transformation are linear but, instead of them, transformations of a higher degree often are needed in practice. If the transformation — or the "map projection" — must be conformal the formulas read

$$\begin{aligned} X &= A + Cx + Dy + E(x^2 - y^2) + F(2xy) + G(x^3 - 3xy^2) + H(3x^2y - y^3) + \dots \\ Y &= B + Cy - Dx + E(2xy) - F(x^2 - y^2) + G(3x^2y - y^3) - H(x^3 - 3xy^2) + \dots \end{aligned}$$

$$(16.30)$$

In some conventional photogrammetric adjustments, the transformation is not conformal. Each of the model coordinates X, Y and Z is independently adjusted by a polynomial of higher degree like

$$X = A + Cx + Dy + Ex^2 + Fxy + Gy^2 + Hx^3 + \dots \qquad (16.31)$$

The reliability of the terms already computed and the need of additional terms can be investigated by the aid of the residuals after tentative adjustments.

16.7 *Adjustment of traverse nets*

The traverse nets are seldom adjusted rigorously by the method of least squares because the observations of an inferior order do not justify the very complicated computations involved by the rigorous adjustment. However, if only the programs are readily available, the automatic computers can be used for a rigorous adjustment with insignificant extra costs.

For computers with a sufficient capacity, the programs for the adjustment of triangulation by coordinates could be used; cfr. Chapter 9.4. For computers with a smaller capacity, or for adjustment of very large nets, the special method of fictitious observations is necessary as explained below.

The original observations consist of angles α and sides S. At fork points where three or more sides meet the angles have usually been measured by series method; in this case, we consider each direction as one observation, but at the *angle points* where only two sides meet the angle itself, or the difference of two observed directions, is considered as one observation. Such an angle can usually be taken as the weight unit. The directions then have weight $p = 2$ or the inverse weight $q = 0.5$. The weights of the observed sides must be estimated as explained in Chapter 5.2. In the numerical example below, the sides have been measured by a tape and we shall use the formula

$$q_s = cs$$

where the constant c has been estimated as follows.

Supposing that the mean error of one observed angle (with weight $p = 1$) is $m = \pm 15$ centesimal seconds and the mean error of the side which is 1 km long is $\mu = \pm 40$ mm/ $\sqrt{\text{km}}$ we obtain

$$m_s^2 = \mu^2 s$$

where s is expressed in kilometers. In our computationformulas, we express differential angles in microradians. Then we have

$$q_s = m_s^2 (\frac{\rho}{m})^2 \qquad \rho = 0.63662$$

or

$$c = \left(\frac{\rho\mu}{m}\right)^2 = 2.88$$

which is rounded off to $c = 3$ kilometers.

The fictitious observations consist of three computed quantities for each chain. A *chain* means here such a succession of sides inside of which there are angle points only but each end point of a chain must be a fork point or a fixed point. All original observations must be grouped to chains, cfr. Tables I and II. The grouping of angles and sides is selfevident, each direction at the fork points must be assigned to the pertinent chain using minus sign at the end point according to the direction in which the first computation of the chain will be performed. If the fork point is a fixed point, the observed direction to the neighbouring fixed point forms alone a "chain" without any measured sides, *e.g.* Chain 1 in Table I.

As unknown parameters of the adjustment serve the coordinates x and y of all new points and the zero directions z at all fork points. The method of fictitious observations means now that the parameters of all new angle points are eliminated before the net adjustment. In this way, the number of normal equations can be essentially reduced.

Anyway, the approximate values for all parameters must be computed first, cfr. Table II. Starting from the nearest fixed point, the computations are carried out by the usual formulas

$$t_i = t_{i-1} + \alpha_i \pm 200^g$$

$$\Delta x_i = s_i \cos t_i$$

$$\Delta y_i = s_i \sin t_i$$

$$x_{i+1} = x_i + \Delta x_i$$

$$y_{i+1} = y_i + \Delta y_i$$

The new fork points are called *nodes*. For each node, at least three different values of each parameter are obtained; one of these values must be choosen to serve as the constant reference value. The differences reference value minus the value computed along any chain serve as constant terms ℓ in the fictitious observation equations:

$$\ell_Z = z'_E - z_E$$

$$\ell_X = x'_E - x_E$$

$$\ell_Y = y'_E - y_E$$

where subscript E means the end point of the chain, prime the reference value.

The fictitious observations can be expressed by formulas

$$Z = [\alpha] \pm n \cdot 200^g = z_E - z'_S$$

$$X = [s \cos t] = x_E - x'_S$$

$$Y = [s \sin t] = y_E - y'_S$$

where the brackets mean a summation over the chain and subscript S means the starting point of the chain. Denoting

$$\xi_i = x_i - x_E$$

$$\eta_i = y_i - y_E$$

we can write the differential formulas

$$v_Z = [v_\alpha]$$

$$v_X = [v_s \cos t] + [v_\alpha \eta]$$

$$v_Y = [v_s \sin t] - [v_\alpha \xi]$$

Using inverse weights q for angles v_α and $q_s = cs$ for sides v_s we compute the weight coefficients, cfr. Tables II and IV.

$$Q_{ZZ} = [q]$$

$$Q_{ZX} = [q\eta]$$

$$Q_{ZY} = [-q\xi]$$

$$Q_{XX} = [q\eta^2] + c[\Delta x \cos t]$$

$$Q_{XY} = [-q\xi\eta] + c[\Delta x \sin t]$$

$$Q_{YY} = [q\xi^2] + c[\Delta y \sin t]$$

In regular cases, each chain gives three fictitious observation equations

$$\delta z_E - \delta z_S + \ell_Z = \upsilon_Z$$

$$-\eta_S \delta z_S + \delta x_E - \delta x_S + \ell_X = \upsilon_X$$

$$\xi_S \delta z_S + \delta y_E - \delta y_S + \ell_Y = \upsilon_Y$$

which are correlated. In order to deal with these correlations, the symmetrical matrix $\mathbf{Q}_c$ of each chain must be inverted to another symmetrical matrix $\mathbf{P}_c = \mathbf{Q}_c^{-1}$ with elements p_{ZZ}, p_{ZX} etc. as shown in Table IV. When the observation equations for all chains have been written, the coefficients form a large matrix **A** and the differences ℓ a vector **1** as shown in Table V but the matrices P_c must be considered as parts of a slanting chain matrix P.

If the starting point of a chain is fixed, there will be no corrections δ_{x_S} and δ_{y_S}. If such a point is an angle point, there will be no δ_{z_S}, but if it is a fork point, there will be an additional "chain" with one equation $\delta_{z_E} = \upsilon_Z$ only.

The normal equations can now be formed by the matrix formulas

$$\mathbf{N} = \mathbf{A}^T\mathbf{P}\ \mathbf{A}$$

$$\mathbf{u} = \mathbf{A}^T\mathbf{P}\ \mathbf{1}$$

as shown in Table VI. The solution of normal equations gives the corrections $\delta\mathbf{z}$, $\delta\mathbf{x}$ and $\delta\mathbf{y}$ for each node. Then the fictitious residuals υ_Z, υ_X, υ_Y can be computed for each chain. Furthermore, we must compute the elements of matrix product **Pv** or

$$V_Z = p_{ZZ}\upsilon_Z + p_{ZX}\upsilon_X + p_{ZY}\upsilon_Y$$

$$V_X = p_{ZX}\upsilon_Z + p_{XX}\upsilon_X + p_{XY}\upsilon_Y$$

$$V_Y = p_{ZY}\upsilon_Z + p_{XY}\upsilon_X + p_{YY}\upsilon_Y$$

If the unknown parameters are eliminated from observation equations (Table V) we obtain condition equations $\mathbf{B}^T\mathbf{v} + \mathbf{w} = \mathbf{o}$ as explained in Chapter 13.2. Matrix **B** is given in Table VII.

In this case, the normal equations can be formed by formula

$$M = B^T Q\, B$$

and it is not necessary to invert the original matrix **Q**. In fact, the same normal equations could be formed from the original observations without using the fictitious observations at all. Anyway, the solution of the normal equations gives now the correlates k by the aid of which we can compute first **Pv** = **Bk** and then **v** = **Q Pv**.

Notwithstanding whether observation equations or condition equation have been used, the residuals of the original observations can be computed for each chain from the elements of **Pv** by formulas

$$v_\alpha = \rho q(V_Z + \eta V_X - \xi V_y)$$

$$v_s = c(\Delta x V_x + \Delta y V_y)$$

Eventually, there are several well known ways to compute the adjusted coordinates for the angle points of each chain.

As an example, a small part of the city traverse net of Helsinki (Fig.13) has been computed by both methods

The weighted square sums of the residuals are

$$[pvv] = 2540 \text{ for angles}$$
$$3798 \text{ for sides}$$
$$6338 \text{ in total}$$

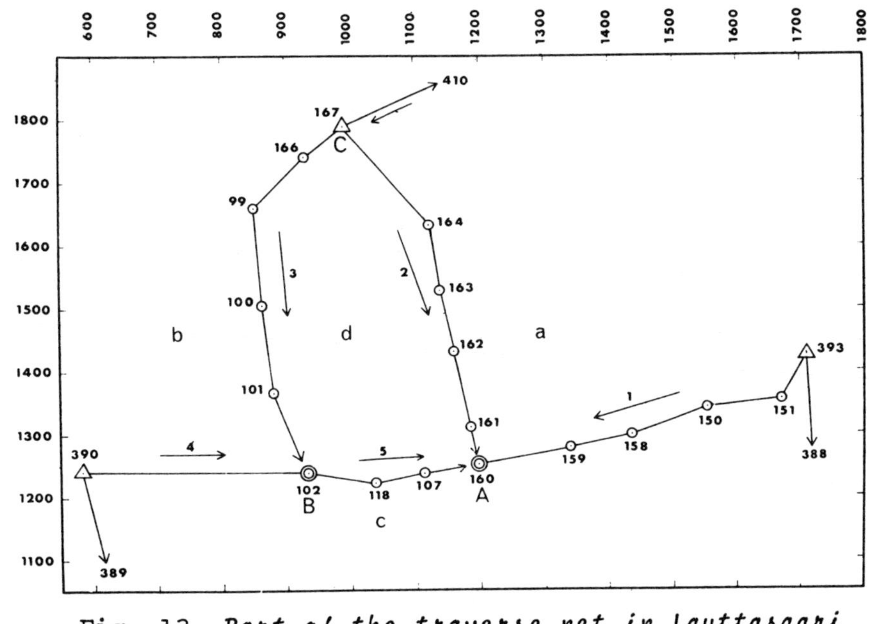

Fig. 13. *Part of the traverse net in Lauttasaari*

The mean error of the unit weight is $\sqrt{6338/9} = \pm 26.5$ microradians = $\pm 16''.9$. The elements of the inverse matrix in Table VI show that, for example, at node A = point 160 the mean errors are $m_z = 15''.6$, $m_x = 8.7$ and $m_y = 13.6$ millimeters. According to our formula (3.70), these figures correspond to class numbers $M = 3.3$ with respect of angle measurements and $M = 3.5$ with respect of coordinate determinations.

Table I *The observed angles and the approximate directions*

Chain	Point	α	q	t	$\cos t$	$\sin t$	v_α	v_t
1	410			$273^g.51456$				
	C=167	-0"	0.5	73.51456	= reference value		-1".05	-1".05
2	167	$81^g.6606$	0.5	155.17516	-.7621 8998	.6473 5341	-2.80	-3.85
	164	235.4292	1	190.60436	-.9891 2889	.1470 5117	0.57	-3.28
	163	196.0722	1	186.67656	-.9781 7989	.2077 5968	3.80	0.52
	162	199.8739	1	186.55046	-.9777 6646	.2096 9683	6.83	7.35
	161	199.6263	1	186.17676	-.9765 1868	.2154 3273	10.65	18.00
	A=160	-95.9914	0.5	290.18536			6.27	24.27
3	167	183.8486	0.5	257.36316	-.6207 8054	-.7839 8439	1.76	0.71
	166	191.2107	1	248.57386	-.7227 6847	-.6910 9025	3.55	4.26
	99	146.2235	1	194.79736	-.9966 6254	.0816 3194	3.82	8.08
	100	197.8228	1	192.62016	-.9932 8854	.1156 6281	5.65	13.73
	101	182.1793	1	174.79946	-.9226 6947	.3855 9182	7.34	21.08
	B=102	-74.4601	0.5	300.33936			4.60	25.67
4	389			386.54438				
	390	313.7965	1	100.34088	-.0053 5451	.9999 8567	12.65	12.65
	B=102	- 0	0.5	300.34088	= reference value		-2.18	10.47
5	102	209.8977	0.5	110.23858	-.1601 3483	.9870 9515	2.41	12.89
	118	176.3422	1	86.58078	.2092 3112	.9778 6621	0.65	13.54
	107	203.6062	1	90.18698	.1535 3288	.9881 4354	-2.45	11.09
	A=160	- 0	0.5	290.18698			-3.02	8.07
6	388			395.65909				
	393	36.7369	1	232.39599	-.8732 9314	-.4871 9512	14.30	14.30
	151	261.0241	1	293.42009	-.1031 7306	-.9946 6342	10.62	24.92
	150	184.1765	1	277.59659	-.3446.9320	-.9387 1540	6.80	31.73
	158	209.0776	1	286.67419	-.2077 9610	-.9781 7217	1.95	33.68
	159	200.8397	1	287.51389	-.1948 7632	-.9808 2782	-1.47	32.21
	A=160	-197.3290	0.5	290.18489	= reference value		-3.24	28.97

[pvv] 1029.50 sec^2

Table II *The observed sides and the approximate coordinates*

Chain	Point	s	Δx or x	Δy or y	η	$-\xi$	pv_s	v_s
2	<u>167</u>		<u>1789.234</u>	<u>988.887</u>				
	167	206.190	-157.155 95	+133.477 80	-.2085	-.5383	14.89	9.21
	164	105.819	-104.668 63	15.560 81	- 750	-.3811	-12.24	-3.88
	163	98.158	- 96.016 18	20.393 27	- 595	- 2764	- 9.26	-2.73
	162	123.638	-120.889 09	25.926 50	- 391	- 1804	- 9.17	-3.40
	161	60.955	- 59.523 70	13.131 70	- 131	- 595	- 8.88	-1.62
	A= 160		1250.980 45	1197.377 08	0	0		
3	167	78.240	- 48.569 87	- 61.338 94	570	- 5504	-21.75	-5.10
	166	122.176	- 81.077 28	- 77.523 74	- 43	- 5018	-21.43	-7.21
	99	155.113	-154.595 32	12.662 18	- 819	- 4207	-11.40	-5.30
	100	140.355	-139.413 01	16.233 85	- 692	- 2661	-10.76	-4.53
	101	137.347	-126.725 88	52.959 88	- 530	- 1267	- 5.11	-2.11
	B= 102		1238.852 64	931.880 23	0	0		
4	<u>390</u>		<u>1240.742</u>	<u>583.763</u>				
	390	348.127	- 1.864 05	348.122 01	- 3481	- 19	- 6.98	-7.28
	B= 102	ref.=	1238.877 95	931.885 01				
5	102	106.742	- 17.093 11	105.364 51	- 2655	121	20.26	6.49
	118	75.628	+ 15.823 73	73.954 07	- 1601	292	- 3.43	-0.78
	107	87.211	+ 13.389 76	86.176 99	- 862	134	0.23	0.06
	A= 160		1250.998 33	1197.380 58	0	0		
6	<u>393</u>		<u>1426.232</u>	<u>1709.436</u>				
	393	82.138	- 71.730 55	- 40.017 23	5120	- 1753	-11.93	-2.94
	151	119.328	- 12.311 43	-118.691 20	4720	- 1035	+50.78	18.18
	150	124.678	- 42.975 66	-117.037 16	3533	- 912	36.86	13.79
	158	96.986	- 20.153 31	- 94.869 01	2362	- 482	45.18	13.15
	159	144.137	- 28.088 89	-141.373 58	1414	- 281	45.91	19.85
	A= 160	ref. =	1250.972 16	1197.447 82	0	0		
							[pvv]	3797.51

Misclosures in x and y:

		x	y
a =	Chains 6 - 1 - 2:	- 8.29	+70.74 mm
b	1 + 3 - 4:	-25.31	- 4.78
c	4 + 5 - 6:	+26.17	-67.24

Table III *Computation of the weight coefficients*

Chain	$[q]$	$[q\eta]$	$[q\xi]$	$[q\eta^2]$	$[q\xi\eta]$	$[q\xi^2]$	$\Delta x\cos t$	$\Delta x\sin t$	$\Delta y\sin t$
1	0.5								
2	5.0	-0.2909	-1.1665	0.0326	0.1090	0.4026	0.4936	-0.1752	0.1012
3	5.0	-0.1799	-1.5905	0.0159	0.0461	0.6671	0.4982	0.0165	0.1250
4	1.5	-0.3481	-0.0019	0.1212	0.0007	0	0	-0.0019	0.3481
5	3.0	-0.3790	0.0486	0.0683	-.0074	0.0011	0.0081	0.0118	0.2615
6	5.5	1.7149	-0.4463	0.6855	-.1862	0.0529	0.0884	0.1348	0.4789

Table IV *The weight coefficients and the weight matrices*

Chain	Q_c			P_c			det Q_c
1	0.5			2.000			
2	5.0	-0.2909	-1.1665	0.432	0.334	0.910	
		1.5133	-0.4168		1.047	1.169	
			0.7062			3.609	2.0732
3	5.0	-0.1799	-1.5905	0.389	0.009	0.593	
		1.5106	0.0956		0.666	-0.048	
			1.0421			1.868	4.0252
4	1.5	-0.3481	-0.0019	2.000	5.746	0.031	
		0.1212	-0.0049		24.759	0.127	
			1.0444			0.958	0.0633
5	3.0	-0.3790	0.0486	0.707	2.939	-0.149	
		0.0926	0.0281		23.134	-1.009	
			0.7855			1.318	0.1016
6	5.5	1.7149	-0.4463	0.522	-1.011	0.304	
		0.9507	0.2182		3.048	-0.749	
			1.4895			0.872	2.6230

Table V *The fictitious observation equations*

Chain		C_z	B_z	B_x	B_y	A_z	A_x	A_y	ℓ	$P\ell$
1	Z	1								
2	Z	-1				1			- 7.38	58.42
	X	0.208					1		- 8.29	71.55
	Y	0.538						1	70.74	238.89
3	Z	-1	1						23.88	12.35
	X	-0.057		1					25.31	16.84
	Y	0.550			1				4.78	21.88
4	Z		1							
	X			1						
	Y				1					
5	Z		-1			1			-32.83	-110.14
	X		0.265	-1			1		-26.17	-769.75
	Y		-0.012		-1			1	67.24	119.92
6	Z					1				
	X						1			
	Y	Matrix **A**						1		

Table VI *The normal equations*

		C_z	B_z	B_x	B_y	A_z	A_x	A_y	u
C	z	2.975	-0.063	-0.073	0.438	0.127	0.514	1.276	83.82
B	z		3.169	2.540	0.759	0.075	3.215	-0.135	-83.33
	x			48.559	-0.930	-2.939	-23.134	1.009	786.59
	y				4.144	0.149	1.009	-1.318	-98.04
A	z					1.661	2.262	1.065	-51.72
	x						27.229	-0.589	-698.20
	y	Matrix **N**						5.799	358.81
								[pℓℓ]	48524.2
		The inverse matrix and the solution							δ
C	z	0.406	0.065	-0.012	-0.092	0.074	-0.031	-0.123	- 1.64
B	z		0.591	- 107	- 110	25	- 161	- 28	16.46
	x			0.055	17	28	57	- 5	- 6.88
	y				0.304	- 89	27	103	- 7.33
A	z					0.855	- 53	- 203	45.45
	x						0.109	20	12.89
	y	Matrix N^{-1}						0.262	-68.64
								[pvv]	6339.7

Table VII *The condition equations*

Chain		a_z	a_x	a_y	b_z	b_x	b_y	c_z	c_x	c_y	Pv = V	v
1	Z	-1	0.208	0.538	1	0.057	-0.550				- 3.30	- 1.65
2	Z	-1									19.69	39.76
	X		-1								19.14	4.26
	Y			-1							45.53	1.21
3	Z				1						14.44	41.98
	X					1					12.87	18.52
	Y						1				17.54	- 3.45
4	Z				-1			1	-0.265	0.012	- 6.85	16.45
	X					-1			1		-76.74	- 6.88
	Y						-1			1	- 7.39	- 7.32
5	Z							1			- 9.49	- 3.78
	X								1		-63.87	- 2.03
	Y									1	10.16	5.72
6	Z	1						-1			-10.19	45.50
	X		1						-1		44.73	12.89
	Y			1						-1	-55.69	-68.64
		Matrix B								[pvv]	6337.9	

Table VIII *The normal equations*

		a_z	a_x	a_y	b_z	b_x	b_y	c_z	c_x	c_y	**w**
a	z	11.000	1.320	-1.882	-0.500	-0.028	0.275	-5.500	-1.715	0.446	- 7.38
	x		2.486	- 142	104	6	- 57	-1.715	- 951	- 218	- 8.29
	y			2.341	269	15	- 148	446	- 218	-1.490	70.74
b	z				7.000	- 500	-1.868	-1.500	746	- 16	-23.88
	x					1.633	75	348	- 214	9	-25.31
	y						2.238	2	4	-1.044	- 4.78
c	z							10.000	590	- 381	32.83
	x								1.455	233	26.17
	y	Matrix **M**								3.320	-67.24
				Solution by the method of Cholesky							**e**
a	z	3.317	0.398	-0.567	-0.151	-0.009	0.083	-1.658	-0.517	0.135	- 2.23
	x		1.526	55	108	6	- 59	- 692	- 488	- 178	- 4.85
	y			1.420	125	7	- 69	- 322	- 342	- 988	49.12
b	z				2.636	- 190	- 698	- 620	+ 290	56	-11.32
	x					1.264	- 45	+ 176	- 125	23	-22.01
	y						1.317	- 265	145	- 831	- 7.89
c	z							2.486	- 191	- 317	14.62
	x								0.822	- 92	53.42
	y	Matrix **K**								1.220	-12.40
									[ee] =	[pvv]	6337.3
	k	-19.69	-19.14	-45.53	14.44	12.87	17.54	- 9.49	-63.87	10.16	6337.8

17 *PHOTOGRAMMETRIC APPLICATIONS*

17.1 *Definitions and notations*

The observed image coordinates are denoted by x, y. Generally, two subscripts are required, one indicating the number of the point, another indicating the number of the photograph. The coordinates are considered uncorrelated and equally weighted. It is supposed the interior orientation of the photographs to be known and the corrections of image coordinates, caused by the lens deformation and the atmospheric refraction, to have already been added. In this chapter, we are not discussing the adjustments which might be necessary for the determination of the parameters involved by these corrections.

The ground coordinates in a three-dimensional Cartesian system are denoted by X, Y, Z. The XY-plane is supposed to be tangent to the reference ellipsoid at some point which is chosen near to the centre of the block to be adjusted. In the practice, however, some grid coordinates and heights above the sea level are used; in that case, corresponding corrections should be made to the image coordinates x, y. The ground coordinates which have been determined by geodetic operations are usually supposed to be errorless; nothing prevents, however, to assign special weights to them as mentioned in Chapter 15.3.

For each photograph, six unknown parameters are needed for the exterior orientation. Three of these consist of ground coordinates X_N, Y_N, Z_N (or H) of the exposure station. The three other parameters refer to the direction of the axes of image coordinates. For exact computations, these directions should be taken in a definite order.

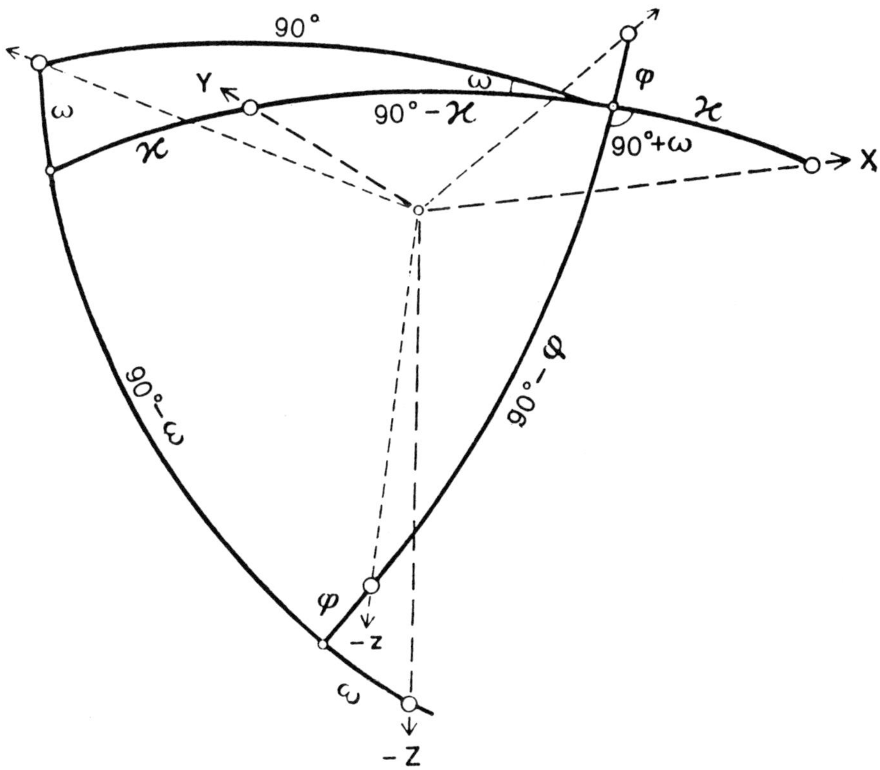

Here the system of fig. 14 is used:

1. X-axis is turned towards Y-axis by angle κ.

2. The new direction of Y-axis is turned towards Z-axis by angle ω.

3. The new direction of X-axis is turned towards the new direction of Z-axis by angle ϕ.

The final direction of X-axis is now parallel to **x**-axis which is supposed to indicate approximately the direction of the flight.

17.2 *Transformation formulas*

According to fig. 14, formula (3.50) gives easily the cosines of all angles between two axes belonging to different systems:

$$
\begin{aligned}
a &= \cos(X,x) &&= +\ \sin\phi\ \sin\omega\ \sin\kappa + \cos\phi\ \cos\kappa \\
b &= \cos(X,y) &&= -\ \cos\omega\ \sin\kappa \\
c &= \cos(X,-z) &&= -\ \cos\phi\ \sin\omega\ \sin\kappa + \sin\phi\ \cos\kappa \\
a' &= \cos(Y,x) &&= -\ \sin\phi\ \sin\omega\ \cos\kappa + \cos\phi\ \sin\kappa \\
b' &= \cos(Y,y) &&= \cos\omega\ \cos\kappa \\
c' &= \cos(Y,-z) &&= +\ \cos\phi\ \sin\omega\ \cos\kappa + \sin\phi\ \sin\kappa \\
a'' &= \cos(-Z,x) &&= -\ \sin\phi\ \cos\omega \\
b'' &= \cos(-Z,y) &&= -\ \sin\omega \\
c'' &= \cos(Z,z) &&= +\ \cos\ \cos\omega
\end{aligned}
\qquad (17.01)
$$

Usually the camera is almost vertical; then the angles ϕ and ω are small and the following formulas give a sufficient accuracy for several purposes:

$$
\begin{aligned}
a &= b' = \cos\kappa \\
-b &= a' = \sin\kappa \\
c &= \phi\cos\kappa - \omega\sin\kappa \\
c' &= \phi\sin\kappa + \omega\cos\kappa \\
a'' &= -\ \phi \\
b'' &= -\ \omega \\
c'' &= 1
\end{aligned}
\qquad (17.02)
$$

In the system of image coordinates, $z = f$ = constant. Using the matrix formula (14.27) the rigorous formulas of transformation can be written in form:

$$
\begin{Bmatrix} x \\ y \\ f \end{Bmatrix} K = \begin{Bmatrix} a & a' & a'' \\ b & b' & b'' \\ c & c' & c'' \end{Bmatrix} \begin{Bmatrix} X - X_N \\ Y - Y_N \\ H - Z \end{Bmatrix} \qquad (17.03)
$$

If the parameters of the exterior orientation and the ground coordinates are known, the last row of (17.03) gives first K and the former rows then the image coordinates.

The following inverted formula can be used if the parameters of exterior orientation, the image coordinates x, y and the third ground coordinate Z are known:

$$\begin{Bmatrix} X - X_N \\ Y - Y_N \\ H - Z \end{Bmatrix} k = \begin{Bmatrix} a & b & c \\ a' & b' & c' \\ a'' & b'' & c'' \end{Bmatrix} \begin{Bmatrix} x \\ y \\ f \end{Bmatrix} \tag{17.04}$$

Here $k = 1/K$ must be computed first. If the flight altitude does not vary very much between the successive exposures and if the height differences of the terrain are small, an approximately constant value of the scale

$$k_o = \frac{f}{H_o} \tag{17.05}$$

can be used. Here $H_o = H - Z_o$ when Z_o is the mean height of the terrain.

When the approximate formulas (17.02) are inserted into (17.03), the latter can be written in form

$$\begin{aligned} fK &= H - Z + (X-X_N)(\phi\cos\kappa - \omega\sin\kappa) + (Y-Y_N)(\omega\cos\kappa + \phi\sin\kappa) \\ xK &= (X-X_N)\cos\kappa + (Y-Y_N)\sin\kappa - (H-Z)\phi \\ yK &= (Y-Y_N)\cos\kappa - (X-X_N)\sin\kappa - (H-Z)\omega \end{aligned} \tag{17.06}$$

When Z/H is small and $Z' = Z - Z_o$, the elimination of K gives

$$\begin{aligned} (X-X_N)k_o\cos\kappa + (Y-Y_N)k_o\sin\kappa &= x + f\phi + \frac{x^2\phi}{f} + \frac{xy\omega}{f} - x\frac{Z'}{H_o} \\ (Y-Y_N)k_o\cos\kappa - (X-X_N)k_o\sin\kappa &= y + f\omega + \frac{xy\phi}{f} + \frac{y^2\omega}{f} - y\frac{Z'}{H_o} \end{aligned} \tag{17.07}$$

If the ground coordinates are given, following approximate transformation formulas give the image coordinates:

$$\begin{aligned} x &= (X-X_N)k_o\cos\kappa + (Y-Y_N)k_o\sin\kappa - f\phi \\ &\quad + x\left(\frac{Z'}{H_o} - x\frac{\phi}{f} - y\frac{\omega}{f}\right) \\ y &= (Y-Y_N)k_o\cos\kappa - (X-X_N)k_o\sin\kappa - f\omega \\ &\quad + y\left(\frac{Z'}{H_o} - x\frac{\phi}{f} - y\frac{\omega}{f}\right) \end{aligned} \tag{17.08}$$

In each formula, the first row gives rough values for x and y, respectively; these values are inserted to the second rows which then give corrections to the rough values.

Differentiation of (17.03) gives, after the elimination of dK:

$$
\begin{aligned}
dx = {} & k\left(a - c\frac{x}{f}\right)(dX - dX_N) \\
& + k\left(a' - c'\frac{x}{f}\right)(dY - dY_N) \\
& - k\left(a'' - c''\frac{x}{f}\right)(dZ - dH) \\
& + \left[y\left(c'' + a''\frac{x}{f}\right) - b''\left(f + x\frac{x}{f}\right)\right]d\kappa \\
& - y\left(\sin\phi + \frac{x}{f}\cos\phi\right)d\omega \\
& - \left(f + x\frac{x}{f}\right)d\phi \\
dy = {} & k\left(b - c\frac{y}{f}\right)(dX - dX_N) \\
& + k\left(b' - c'\frac{y}{f}\right)(dY - dY_N) \\
& - k\left(b'' - c''\frac{y}{f}\right)(dZ - dH) \\
& - \left[x\left(c'' + b''\frac{y}{f}\right) - a''\left(f + y\frac{y}{f}\right)\right]d\kappa \\
& + \left[x\sin\phi - \left(f + y\frac{y}{f}\right)\cos\phi\right]d\omega - x\frac{y}{f}d\phi
\end{aligned}
\tag{17.09}
$$

These rigorous differential formulas should be programmed for the high speed computers for photogrammetric adjustments. For manual computations, approximate formulas are more usual:

$$
\begin{aligned}
dx = {} & k_o\cos\kappa\,(dX - dX_N) + k_o\sin\kappa\,(dY - dY_N) + \frac{x}{H_o}(dZ - dH) \\
& + y\,d\kappa - (f^2 + x^2)\frac{d\phi}{f} - xy\frac{d\omega}{f} \\
dy = {} & k_o\cos\kappa\,(dY - dY_N) - k_o\sin\kappa\,(dX - dX_N) + \frac{y}{H_o}(dZ - dH) \\
& - x\,d\kappa - xy\frac{d\phi}{f} - (f^2 + y^2)\frac{d\omega}{f}
\end{aligned}
\tag{17.10}
$$

17.3 *Absolute orientation of one photograph*

The six parameters of the exterior orientation can be computed if the image coordinates x, y are measured for three points for which the ground coordinates X, Y, Z are known. These points must, however, fulfill the condition that if a circle is drawn through these points and a perpendicular cylinder drawn through the circle the mantle of the cylinder must not go through the exposure station. Therefore, usually such points should be chosen which are situated at different borders of the image.

This *pyramid problem* could be solved by closed formulas. In practice, however, more than three points should be used because there often are rough errors in the coordinates. Then an adjustment is necessary. For the adjustment, an approximate solution of the pyramid problem is sufficient. For this purpose, formulas (17.07) are taken in form

$$\begin{aligned} (X-X_N)\,\alpha + (Y-Y_N)\,\beta &= x\,(1-\frac{Z}{H}) + f\sin\phi + x^2\gamma + xy\delta \\ (Y-Y_N)\,\alpha - (X-X_N)\,\beta &= y\,(1-\frac{Z}{H}) + f\sin\omega + xy\gamma + y^2\delta \end{aligned} \qquad (17.11)$$

where H must be estimated beforehand and as well as possible. Each point gives these two equations. X_N and Y_N are elimated by taking differences between two points; at the same time, the terms $f\sin\phi$ and $f\sin\omega$ disappear too. Two pairs of points give four equations from which the unknown quantities α, β, γ, δ can be solved. Eventually

$$\begin{aligned} H &= \frac{f}{\sqrt{\alpha^2+\beta^2}} \\ \cos\kappa &= \frac{H}{f}\,\alpha \\ \sin\kappa &= \frac{H}{f}\,\beta \\ \sin\phi &= f\gamma \qquad \sin\omega = f\delta \end{aligned} \qquad (17.12)$$

Introducing auxiliary quantities

$$\begin{aligned} A &= x\,(1 - \frac{Z}{H}) + f\,\sin\phi + x^2\gamma + xy\delta \\ B &= y\,(1 - \frac{Z}{H}) + f\,\sin\omega + xy\gamma + y^2\delta \\ C &= \frac{H}{f}\cos\kappa \\ D &= \frac{H}{f}\sin\kappa \end{aligned} \tag{17.13}$$

the other parameters can be computed by formulas

$$\begin{aligned} X_N &= X - AC + BD \\ Y_N &= Y - AD - BC \end{aligned} \tag{17.14}$$

from any point.

These approximate parameters are not always accurate enough for an adjustment, especially when the photographs have been taken from a low altitude above a hilly terrain. Therefore, it is advisable to improve them by differential formulas using the same chosen points before the adjustment of all points. For this purpose, the approximate parameters are inserted in the rigorous formulas (17.01) and (17.03). The resulting image coordinates are denoted by x', y'. Comparison with the measured coordinates reveals small differences

$$\begin{aligned} p &= x' - x \\ q &= y' - y \end{aligned} \tag{17.15}$$

Differential formulas (17.09) or (17.10) to which $dx = -p$, $dy = -q$ and $dX = dY = dZ = 0$ are inserted give now six equations for an improvement of the parameters on the basis of the three chosen points. The computations are renewed on basis of all given points using the improved parameters and inserting

$$\begin{aligned} dx &= v_x - p \\ dy &= v_y - q \end{aligned} \tag{17.16}$$

In this way, the observation equations are obtained. Usually they can be considered equally weighted. Then the normal equations can be formed and solved for the final improvement of the parameters.

In manual computations, it is advisable to simplify the observation equations by the aid of temporary parameters

$$
\begin{aligned}
dx_N &= +\, k_o \cos\kappa \; dX_N + k_o \sin\kappa \; dY_N + \ell\, d\phi \\
dy_N &= -\, k_o \sin\kappa \; dX_N + k_o \cos\kappa \; dY_N + \ell\, d\omega \\
dh &= \frac{dH}{H_o} \\
d\phi' &= \frac{d\phi}{\ell} \\
d\omega' &= \frac{d\omega}{\ell}
\end{aligned}
\tag{17.17}
$$

The observation equations read now

$$
\begin{aligned}
dx_N + x\,dh - y\,d\kappa + x^2 d\phi' + xy\,d\omega' &= p - v_x \\
dy_N + y\,dh + x\,d\kappa + xy\,d\phi' + y^2 d\omega' &= q - v_y
\end{aligned}
\tag{17.18}
$$

Denoting $r^2 = x^2 + y^2$, the coefficients and the known terms of the normal equations can be formed as follows

$$
N = \left\{
\begin{array}{cccccc|c}
dx_N & dy_N & dh & d\kappa & d\phi' & d\omega' & u \\
\underline{n} & & +[x] & -[y] & +[x^2] & +[xy] & -[p] \\
 & \underline{n} & +[y] & +[x] & +[xy] & +[y^2] & -[q] \\
 & & \underline{[r^2]} & & +[xr^2] & +[yr^2] & -[px] - [qy] \\
 & & & \underline{[r^2]} & & & +[py] - [qx] \\
 & & & & \underline{[x^2 r^2]} & +[xyr^2] & -[px^2] - [qxy] \\
 & & & & & \underline{[y^2 r^2]} & -[pxy] - [qy^2] \\
 & & & & & & [p^2 + q^2]
\end{array}
\right\}
\tag{17.19}
$$

After the solution of the normal equations, the corrections of the parameters can by found by equations

$$
\begin{aligned}
dH &= H dh \\
d(\sin\kappa) &= \cos\kappa \; d\kappa \\
d(\cos\kappa) &= -\sin\kappa \; d\kappa \\
d(\sin\phi) &= f \; d\phi' \\
d(\sin\omega) &= f \; d\omega' \\
dX_N &= C\,(dx_N - f^2 d\phi') - D\,(dy_N - f^2 d\omega') \\
dY_N &= D\,(dx_N - f^2 d\phi') + C\,(dy_N - f^2 d\omega')
\end{aligned}
\tag{17.20}
$$

where C and D have the same meaning as in (17.13). We give (17.20) also in the following matrix form; at the same time, we obtain the angles κ, ϕ, ω in seconds of arc, instead of radians

$$
\begin{Bmatrix} dX_N \\ dY_N \\ dH \\ d\kappa'' \\ d\phi'' \\ d\omega'' \end{Bmatrix}
=
\begin{Bmatrix}
\frac{H}{f}\cos\kappa & -\frac{H}{f}\sin\kappa & & & -Hf\cos\kappa & +Hf\sin\kappa \\
\frac{H}{f}\sin\kappa & +\frac{H}{f}\cos\kappa & & & -Hf\sin\kappa & -Hf\cos\kappa \\
 & & H & & & \\
 & & & \rho'' & & \\
 & & & & \rho'' f & \\
 & & & & & \rho'' f
\end{Bmatrix}
\begin{Bmatrix} dx_N \\ dy_N \\ dh \\ d\kappa \\ d\phi' \\ d\omega' \end{Bmatrix}
\tag{17.21}
$$

The square matrix of the coefficients in (17.21) may be denoted by G. Then the weight coefficients of the orientation parameters can be computed by matrix formula

$$
Q = G\,\bar{N}\,G^T \tag{17.22}
$$

As a schematic example, suppose that there are nine control points with equal distances on the photograph. The approximate image coordinates x, y are +1, 0 and -1 (decimeters) in all combinations. The coefficients of the observation equations (17.18) form then the following matrix

Table I

Observation equations of a schematic example

Equation	dx_N	dy_N	dh	$d\kappa$	$d\phi'$	$d\omega'$
1 x	1	0	1	-1	1	1
y	0	1	1	1	1	1
2 x	1	0	1	0	1	0
y	0	1	0	1	0	0
3 x	1	0	1	1	1	-1
y	0	1	-1	1	-1	1
4 x	1	0	0	-1	0	0
y	0	1	1	0	0	1
5 x	1	0	0	0	0	0
y	0	1	0	0	0	0
6 x	1	0	0	1	0	0
y	0	1	-1	0	0	1
7 x	1	0	-1	-1	1	-1
y	0	1	1	-1	-1	1
8 x	1	0	-1	0	1	0
y	0	1	0	-1	0	0
9 x	1	0	-1	1	1	1
y	0	1	-1	-1	1	1

The coefficient matrix of the normal equations is N and that of the reduced equations R as follows:

$$N = \begin{Bmatrix} 9 & & & & 6 & \\ & 9 & & & & 6 \\ & & 12 & & & \\ & & & 12 & & \\ 6 & & & & 10 & \\ & 6 & & & & 10 \end{Bmatrix} \qquad R = \begin{Bmatrix} 9 & & & & 6 & \\ & 9 & & & & 6 \\ & & 12 & & & \\ & & & 12 & & \\ & & & & 6 & \\ & & & & & 6 \end{Bmatrix}$$

With the values $H = 1500$ meters, $f = 150$ mm, $\kappa = 0$, we obtain

$$G = \begin{Bmatrix} 10000 & & & & -22500 & \\ & 10000 & & & & -22500 \\ & & 15000 & & & \\ & & & 206265 & & \\ & & & & 309397 & \\ & & & & & 309397 \end{Bmatrix}$$

Formula (17.22) gives now

$$Q = \begin{Bmatrix} 1.53 & & & & 15.04 & \\ & 1.53 & & & & 15.04 \\ & & 0.19 & & & \\ & & & 35.43 & & \\ 15.04 & & & & 159.54 & \\ & 15.04 & & & & 159.54 \end{Bmatrix} \cdot 10^8$$

If the *standard error* of one image coordinate is $m = \pm 0.004$ mm, the standard errors of the parameters are for

X_N and Y_N: $\pm$ 49 mm
H : 17 mm
κ : 2".4
ϕ and ω : 5".1

Our second example is taken from an actual photogrammetric survey. The approximate flight altitude was 582 meters, f = 152.83 mm, hence K_o = 3800. The plates have been measured by a stereocomparator with *a precision* of 2 microns.

Table II. *Ground and plate coordinates*

Point	X	Y	Z	x	y
164	605.167	841.815	6.443	6.7673	31.4825
208	466.205	765.968	4.694	24.8942	- 6.4467
219	520.649	443.735	1.325	109.5385	3.6734
220	484.171	502.563	1.465	93.7162	- 5.0968
221	540.905	540.876	4.455	85.0842	10.2915
235	439.264	670.610	3.787	49.4999	-14.7438
240	174.230	843.256	1.102	0.6282	-81.3607
241	182.166	911.584	1.000	-17.0842	-78.5053
Unit		meter		millimeter	

Table III.

Approximate values of the orientation parameters

Equations (17.11) and their solution

Equation	α	β	γ	δ	= observed
164 x = 1	605	842	0	- 2	6.70
220 = 2	484	503	-88	+ 5	93.48
241 = 3	182	912	- 3	-13	- 17.05
164 y = 4	842	-605	- 2	-10	31.13
220 = 5	503	-484	+ 5	0	- 5.09
241 = 6	912	-182	-13	-62	- 78.38
1 - 2 = 7	121	339	+88	- 7	- 86.78
4 - 5 = 8	339	-121	- 7	-10	+ 36.22
1 - 3 = 9	423	- 70	+ 3	+11	+ 23.75
4 - 6 = 10	-70	-423	+11	+52	+109.51
7.9 → 11		359	+87	-10	- 93.57
8.9 → 12		- 65	- 9	-19	+ 17.19
10.9 → 13		-435	+11	+54	+113.44
11.13 → 14			+96	+36	+ 0.05
12.13 → 15			-11	-27	+ 0.24
15.14 → 16				-23	+ 0.25
Unit	meter		cm^2		mm
Solution	+13	-262	+0.45	-1.09	
	10^{-6}		m^{-1}		

Formulas (17.12) - (17.14) give then:

$\cos \kappa = +0.0496$ $H = 583$ m

$\sin \phi = +0.0068$ $X_N = 495$

$\sin \omega = -0.0166$ $Y_N = 866$

Table IV. *Iteration by formulas* (17.01)

Step	1.	corr.	2.	corr.	3.	corr.
X_N	495.000	+0.754	495.754	+ 117	495.871	- 5
Y_N	866.000	+1.243	867.243	+ 98	867.341	0
H	583.000	-1.372	581.628	- 48	581.580	- 4
cos κ	0.049 6000	+ 1290	49 7290	- 308	49 6982	- 21
sin κ	- 998 7692		- 998 7627		- 998 7643	
sin ϕ	6 8000	+26900	9 4900	+1615	9 6515	+ 11
cos ϕ	999 9769		999 9550		999 9534	
sin ω	- 16 6000	-10160	- 17 6160	-1932	- 17 8092	+ 60
cos ω	999 8622		999 8448		999 8414	
a	0.049 7116		49 8938		49 8676	
a'	- 998 7405		- 998 7094		- 998 6059	
a''	- 6 7991		- 9 4885		- 9 6500	
b	998 6316		998 6077		998 6059	
b'	49 5932		49 7213		49 6903	
b''	16 6000		17 6160		17 8092	
c	- 16 2419		1 17 1215		- 17 3067	
c'	- 7 6150		- 10 3542		- 10 5246	
c"	999 8391		999 7998		999 7948	

Table V. *Computation of image coordinates* by (17.03) and (17.15)

P	$X - X_N$	$Y - Y_N$	$H - Z$	K	p	q
164	+110.167	- 24.185	576.557	3761.428	+ 68.1	- 8.4
220	- 10.829	-363.437	581.535	3823.758	+ 36.3	+ 79.6
241	-312.834	+ 45.584	582.000	3838.515	+141.5	+224.0
Unit		meter		1	micron	

Table VI. *Corrections by formulas* (17.09)

P	dω	dφ	dκ	dH	dY_N	dx_N	q, p	
Unit	dm			10^{+3}			micron	
241 y	- 1.933	- 0.088	+ 0.156	+ 0.138	- 0.012	- 0.258	+224.0	
x	- 82	- 1.547	- 811	+ 27	+ 260	- 12	+141.5	
220 y	- 1.524	+ 31	- 947	+ 13	- 13	- 261	+ 79.6	
x	+ 32	- 2.103	- 86	- 162	+ 260	- 16	+ 36.3	
164 y	- 1.593	- 14	- 79	- 50	- 14	- 266	- 8.4	
x	- 16	- 1.531	+ 289	- 14	+ 265	- 13	+ 68.1	
Sum	- 5.116	- 5.252	- 1.477	- 48	+ 747	- 827	+541.1	
	Solution							1.step corr.
dω	-1.933	-0.088	+0.156	+0.138	-0.012	-0.258	+224.0	-0.001016
dφ	-.0426	-1.544	- 818	+ 21	+ 261	- 15	+132.0	+ 2690
dκ	-.7883	+.0651	-1.123	- 94	+ 14	- 58	- 88.4	+ 129
dH	+ 164	-1.3622	+.9186	- 276	- 84	- 71	-221.2	-1.372meters
dY_N	-.8241	+ 379	- 2122	- 5196	+ 47	- 5	- 54.3	+1.243 "
dX_N	- 83	- 9915	+ 9783	- 4654	-1.2638	- 27	+ 20.5	+0.754 "
Sum	-2.6470	-3.2518	+0.6847	-1.9851	-2.2642	-1.0000		

Return to table IV.

Table VII. *Observation equations* (17.18)

P	x	y	xy	x^2	y^2	r^2	p 2. step	q 2. step	p 3. step	q 3. step
164	0.068	0.315	0.021	0.005	0.099	0.104	+1.0	-0,6	+0.9	0.0
208	0.249	-0.064	- 16	62	4	66	-3.8	+4.5	-2.7	+2.1
219	1.095	0.037	40	1.200	1	1.201	+5.3	-3.9	+1.1	-3.3
220	0.937	-0.051	- 48	878	3	881	+0.4	-0.6	-2.5	-0.2
221	0.851	0.103	88	724	11	735	+2.4	-1.6	+2.0	-1.9
235	0.495	-0.147	- 73	245	22	267	-3.5	+4.0	-3.1	+2.2
240	0.006	-0.814	- 5	0	662	662	-1.0	+2.0	+0.9	+1.9
241	-0.171	-0.785	134	29	616	645	+0.7	-0.1	+2.6	-2.7
Sum	3.530	-1.406	0.141	3.143	1.418	4.561	+1.5	+3.7	-0.8	-1.9
Unit	dm		dm²				micron			

Table VIII. *Normal equations*, formulas (17.19)

	dx_N	dy_N	dh	$d\kappa$	$d\phi'$	$d\omega'$	2.step	3.step	Unit
	8.000		+3.530	+1.406	+3.143	+0.141	-1.50	+0.80	μ
		8.000	-1.406	+3.530	+0.141	+1.418	-3.70	+1.90	μ
			4.561		+2.816	-0.982	-2.59	+2.21	μ·dm
				4.561			+4.87	+1.85	"
					2.835	+0.135	-6.71	+1.13	$\mu\cdot dm^2$
						0.864	-2.13	-0.60	"
							121.9	71.4	μ^2
Step	Solution								
	-.4413	+.1758	2.755		1.454	-0.795	-2.58	+2.19	μ·dm
	-.1758	-.4413		2.755	- 615	- 650	+6.77	+0.87	"
	-.3929	- 177	-.5278	+.2233	0.694	+ 329	-3.18	-0.18	$\mu\cdot dm^2$
	- 177	-.1772	+ 2886	+ 2361	-.4738	0.072	+0.91	-0.03	"
	+ 0.19	+ 0.46	+ 0.94	- 2.46	+ 4.58	-12.64	74.8		μ^2
2.	+ 0.46	+ 2.42	- 8.29	- 3.08	+10.57	-16.64	74.7		"
	- 0.10	- 0.24	- 0.80	- 0.32	+ 0.26	+ 0.39		68.8	"
3.	+ 0.22	- 0.34	- 0.72	- 0.21	+ 0.07	+ 0.39		68.8	"
					$n - u = 10$		$m^2 = 6.88$		"
	Matrix G by formula (17.20)								
dX_N	189	3801			- 441	- 8878	+ 117	- 5	mm
dY_N	- 3801	189			8878	- 441	+ 98	0	"
dH			581.6				- 48	- 4	"
$d\kappa$				1			- 308	- 21	10^{-7}
$d\phi$					1.5283		+1615	+ 11	"
$d\omega$						1.5283	-1932	+ 60	"

Return to Table IV.

Table IX. *Standard errors*

	$\bar{N}$						m	Unit
x_N	0.38	+0.23	-0.70	-0.30	+0.32	-1.28	± 1.6	μ
y_N		0.51	-0.77	-0.47	+0.57	-1.84	1.9	"
h			4.80	+0.81	-4.31	+7.50	5.8	
κ				0.67	-0.54	+1.82	2.1	
ϕ'					4.57	-6.60	5.6	
ω'						13.93	9.8	
	$G\bar{N}G^T$							
X_N	1181	+536	- 39	- 18	+ 90	- 196	± 90	mm
Y_N		396	- 23	- 4	+ 65	- 92	52	"
H			162	+0.3	- 4	+ 7	33	"
κ				0.67	- 1	+ 3	4.4	1"
ϕ					10.7	- 15	18	"
ω						32.5	31	"

The standard error of one image coordinate, $m = \pm 2.6$ micron, is too small because, for this example, only the best fitting points of the entire survey were selected. The complete adjustment of two photographs gave:

Photograph nr. 50 with 17 points: $m = \pm 8.7$ microns

" 51 " 14 " $m = \pm 8.9$ "

Relative orientation of 50 and 51 with 14 common points (without ground coordinates): $m = \pm 3.9$ microns.

These values indicate that the photogrammetric survey has been more accurate than the geodetic survey of the control points.

17.4 *The relative orientation*

In the routine of photogrammetry, the problem of exterior orientation is solved by special plotting instruments with mechanical or optical trains, instead of the analytical treatment of the measured image coordinates. The first photograph is orientated approximately only and the other photographs of the strip are attached one by one to the same system. Eventually, the absolute orientation of the entire strip can be improved analytically by the aid of correction polynomials like (16.31).

In this chapter, the relative orientation, or the connection of one new photograph to the orientation system of the previous photograph, will be performed by the aid of a numerical adjustment. Suppose that there are n pass points for which the image coordinates

x, y have been measured from both photographs. Initially, it is not necessary to know the ground coordinates X, Y for any of them. The unknown parameters of the problem can be divided in two groups: the orientation elements of the new photograph and the ground coordinates of the pass points. In the first group, we introduce the temporary quantities (17.17), in the second group, for each pass point three other temporary quantities

$$\begin{aligned} dx &= k_o(dX \cos \kappa + dY \sin \kappa) \\ dy &= k_o(dY \cos \kappa - dX \sin \kappa) \\ dz &= dZ/H_o \end{aligned} \tag{17.23}$$

The differential formulas (17.10) give then two equations for each pass point and for each photograph. For the first photograph these equations are short:

$$\begin{aligned} - dx - x_1 dz &= p_1 - v_{x1} \\ - dy - y_1 dz &= q_1 - v_{y1} \end{aligned} \tag{17.24}$$

For the second photograph the equations read

$$\begin{aligned} dx_N - dx + x_2(dh - dz) - y_2 d\kappa + x_2{}^2 d\phi' + x_2 y_2 d\omega' &= p_2 - v_{x2} \\ dy_N - dy + y_2(dh - dz) + x_2 d\kappa + x_2 y_2 d\phi' + y_2{}^2\, d\omega' &= q_2 - v_{y2} \end{aligned} \tag{17.25}$$

From these four equations, dx and dy can be eliminated by subtractions. Then two equations remain for each pass point. When dz is eliminated from these, dh disappears at the same time because y_1 and y_2 are approximately equal and $x_1 - x_2$ is approximately constant (basis b) for all pass points. The relative orientation gives, therefore, only five parameters for the orientation of the new photograph; these can be computed if there are, at least, five pass points. In order to get a complete orientation, some ground coordinates should be known, at least, the height for one point.

In practice, there always are more than five pass points and usually some points, the ground coordinates of which already have been determined in the connection of the orientation of the first photograph. Therefore, (17.25) and (17.24) must be considered as observation equations from which the normal equations can be formed and solved.

As an example, we take the image coordinates of Table I/17.4

for two photographs. For the four first pass points we use the known ground coordinates of Table II/17.3, those of the remaining four points are determined by the adjustment. The orientation of photograph nr. 51 is supposed to be known; here we shall use the results given in Table IV/17.3. The unknown parameters have been determined by successive approximations. For two steps only two old and two new points, for two more steps the adjustment of all eight points have been used.

Table I. *Observed image coordinates*

Photo	Point	x	y	x^2	xy	y^2	r^2
51	1 = 164	6.7673	31.4825	46	213	991	1037
	208	24.8942	- 6.4467	620	- 160	42	661
	240	0.6282	- 81.3607	0	- 51	6619	6620
	2 = 241	- 17.0842	- 78.5053	292	1341	6164	6456
	221	85.0842	10.2915	7239	875	106	7344
	226	101.1921	- 47.2957	10239	-4786	2237	12477
	3 = 237	87.2441	- 97.2298	7611	-8482	9454	17064
	4 = 251	91.5745	67.7237	8385	6201	4586	12971
	Sum	380.3004	-201.3405	34432	-4849	30199	64630
50	1 = 164	- 89.2310	31.6586	7962	-2825	1002	8964
	208	- 70.2929	- 6.3065	4941	443	40	4981
	240	- 93.6925	- 82.3862	8778	7719	6788	15566
	2 = 241	-111.7536	- 79.5711	12488	8892	6351	18819
	221	- 9.6136	10.7936	92	- 104	116	209
	226	8.1545	- 47.1903	66	- 385	2227	2293
	3 = 237	- 5.0715	- 98.1299	26	498	9629	9655
	4 = 251	- 6.5434	68.1743	43	- 446	4647	4690
	Sum	-378.0440	-202.9575	34396	13792	30780	65177

Table II. *Approximations for the parameters*

Step	X_N	Y_N	H	cos κ	sin φ	sin ω
1.	513.9	505.1	581.6	0.0497	0.0000	0.0000
	- 9.6	+ 4.1	+ 0.4	- 44	+ 105	- 46
2.	504.3	509.2	582.0	453	105	- 46
	+ 399	+ 510	+ 290	+ 208	+ 808	- 448
3.	504.699	509.710	582.290	45508	11308	- 5048
	- 1.101	- 684	+ 75	- 704	- 737	+ 14
4.	503.598	509.026	582.365	44804	10571	- 3648
	- 4	+ 48	+ 66	+ 12	+ 93	+ 35
5.	503.594	509.074	582.431	+0.044816	+0.010664	-0.00361
	X	Y	Z	X	Y	Z
		237			251	
1.	126.8	505.3	0.0	760.0	526.6	0.0
	+ 0.2	+ 0.1	+ 0.5	- 9.4	+11.7	+20.1
2.	127.0	505.4	0.5	750.6	538.3	20.1
	- 217	- 116	- 202	+ 187	+ 287	+ 284
3.	126.783	505.284	0.298	750.787	538.587	20.384
	+ 23	+ 34	+ 70	+ 220	- 329	- 572
4.	126.806	505.318	0.368	751.007	538.258	19.812
	- 5	+ 1	0	+ 5	- 15	- 16
5.	126.801	505.319	0.368	751.012	538.243	19.796
		221			226	
3.	541.127	541.157	4.937	323.906	463.791	1.423
	- 232	- 207	- 377	- 321	- 89	- 139
4.	540.895	540.950	4.560	323.585	463.702	1.284
	+ 3	- 11	- 10	- 7	- 9	- 3
5. ✓	540.898	540.939	4.550	323.578	463.693	1.281

Table III. *Computed image coordinates*

Step		1.		2.		3.		4.		5.	
Photo	Point	p	q	p	q	p	q	p	q	p	q
51	221					4.2	67.2	- 1.5	-1.7	-0.1	- 1.1
	226					4.8	76.0	- 1.6	3.8	0	2.2
	237	-0.05	0.04	2.4	24.7	-1.5	2.1	0.6	-3.4	0.3	- 4.7
	251	0	-0.01	28.1	-85.1	0.7	3.4	- 1.6	-7.5	-0.1	- 8.2
50	164	1.09	-2.99	- 6.1	22.0	3.6	- 1.9	- 1.4	-2.9	3.5	- 8.9
	208					26.9	-42.1	3.5	-2.3	6.8	- 5.9
	240					6.2	0.5	- 6.4	3.5	-5.4	3.0
	241	0.54	-2.32	- 4.6	7.7	1.2	0.7	- 4.2	0.6	-3.8	- 0.7
	221					0.1	0.1	- 2.8	6.6	-0.2	1.2
	226					0	- 0.1	0.7	-0.7	-0.2	- 2.2
	237	-0.04	-3.45	-18.5	- 8.8	-2.3	2.1	2.4	1.8	-0.4	4.4
	251	4.11	-3.31	54.4	-26.3	2.0	- 2.6	- 4.5	22.6	0.1	8.2
	Unit	mm		micron		micron		micron		micron	
50	Sum					37.7	-43.3	-12.7	29.2	0.3	- 0.9
50+51	$[p^2+q^2]$						12925		784		414
								$n - u = 6$		m^2	69
										m	± 8.3

17.5 *Block adjustments*

The mutual orientation of two successive photographs is always useful for the investigation and elimination of the rough errors in the image and ground coordinates. The full exploitation of the possibilities of analytical photogrammetry, however, requires an extensive block adjustment by the rigorous methods of least squares. If each point is visible on three different photographs in the line of flight and on three overlapping strips or, in total on nine different photographs, the coordinates obtained by a block adjustment are as good as those obtained by secondary triangulations. The first order ground control is necessary only for a few points at the corners of the block. Nevertheless, the geometric levelling of the heights of the targets usually can be carried out without significant extra costs; in this way, the accuracy of the photogrammetric horizontal coordinates is increased and the demand on the extensive computations decreased.

In a block adjustment, the number of parameters is necessarily very large: six for each photograph and three for each new point which has no geodetic coordinates(two if the target has a levelled height). The solution of the normal equations, however, is nowadays possible by the aid of large electronic computers.

Each pair of the observation equations contains nine (or eight) parameters as shown by (17.24). The normal equations can be divided in two groups. Each equation of the first group contains six parameters of the orientation and a greater number of unknown coordinates. Each equation of the second group contains three (or two) unknown coordinates and a very great number of orientation parameters. Denoting by brackets a summation over all n points visible on one photograph P and by parentheses a summation over all m photographs on which one point A appears, the normal equations obtained from the simplified formulas (17.24) can be written in form of Table I in which the first six equations for one photograph P involve more points (A, B, C etc.) and the three last equations for one point A involve more photographs (P, Q etc.).

Table I. *General form of normal equations*

dx_P	dy_P	dH'_P	$d\kappa_P$	$d\phi'_P$	$d\omega'_P$	dX'_A	dY'_A	dZ'_A	Constant term
n		$+[x]$	$-[y]$	$+[x^2]$	$+[xy]$	-1		$-x_{AP}$	$-[p]$
	n	$+[y]$	$+[x]$	$+[xy]$	$+[y^2]$		-1	$-y_{AP}$	$-[q]$
		$[r^2]$		$+[xr^2]$	$+[yr^2]$	$-x_{AP}$	$-y_{AP}$	$-r^2_{AP}$	$-[px+qy]$
			$[r^2]$			$+y_{AP}$	$-x_{AP}$		$[py-qx]$
				$[x^2r^2]$	$+[xyr^2]$	$-x^2_{AP}$	$-xy_{AP}$	$-xr^2_{AP}$	$-[px^2+qxy]$
					$[y^2r^2]$	$-xy_{AP}$	$-y^2_{AP}$	$-yr^2_{AP}$	$-[pxy+qy^2]$
						m		$+(x)$	(p)
							m	$+(y)$	(q)
								(r^2)	$(px+qy)$

The orientation parameters are eliminated first because they are of minor importance. Denoting by N the matrix of the remaining normal equations, the inverse matrix $\bar{N}$ gives the weight coefficients for the ground coordinates of the new points.

The method has been tested several times in Finland with extraordinary success. The standard error of unit weight (one image coordinate) has always been about 4 microns. The weight coefficients of the new ground coordinates are, of course, very variable according to the number of the geodetic points and to the flight altitude. Therefore, only two schematic cases will be given here as examples.

1. *Strip adjustment.* There are 21 points situated as shown by the following scheme:

Photograph nr		1		3		5	
Point nr	1	4	7	10	13	16	19
	2	5	8	11	14	17	20
	3	6	9	12	15	18	21
Photograph nr			2		4		

Points nr 1 - 6 and 16 - 21 are supposed to be geodetic control points, points nr 7-15 new points. There are five photographs with nine points on each. The observation equations in form (17.24) contain only coefficients +1, 0 and -1 like in Table I/17.3. The matrix of the 57 normal equations (30 for the orientation, 27 for the new points) is too large to be printed here. From the inverted matrix, some small parts are shown below:

Table II. *Weight coefficients for some points*

Point	x	y	z
10	1.18	0 2.90	0 1.81 3.29
11	1.02	0 1.13	0 0 2.68
13	0.93	0.17 2.30	0.25 1.52 2.30
14	0.74	0 0.85	0.23 0 1.81

Table III. *Weight coefficients for photographs*

Photograph	x_N	y_N	h	κ	ϕ'	ω'
1	0.22	0	-0.03	0	-0.10	0
		0.34	0	-0.11	0	-0.24
			0.29	0	-0.23	0
				0.16	0	+0.09
					0.43	0
						0.31
3	1.10	0	0	0	-0.23	0
		0.77	0	0	0	-0.43
			1.13	0	0	0
				0.24	0	0
					0.51	0
						0.62

Suppose that m = 4 microns, f = 150 mm, H = 1500 meters, the distances between neighboring points on the image 100 mm. Then the standard errors of the ground coordinates of point 11 are:

along the strip	± 40 mm
across the strip	43
in the height	98

If, however, the heights have been determined by a geometric leveling the corresponding results are:

Table IV. *Weight coefficients for points with fixed heights*

Point	10		11		13		14	
x	1.07	0	0.92	0	0.84	0.05	0.69	0
y		0.89		0.66		0.80		0.57

Standard errors of point 11:

along the strip	± 38 mm
across the strip	32

2. *Block adjustment.* In the square block shown at the right the control points with complete geodetic coordinates are denoted by a cross and the other points with known heights only by a circle. As above, there are nine points on each photograph. The nine points in the central part of the block are visible on nine photographs each. The total number of photographs is 25, that of the new points 45.

Square block

o	o	o	o	o	o	o
o	x	o	o	o	x	o
o	o	o	o	o	o	o
o	o	o	o	o	o	o
o	o	o	o	o	o	o
o	x	o	o	o	x	o
o	o	o	o	o	o	o

The weight coefficients of the coordinates turn out to be almost constant for all new points in the central part of the block; the minimum value at the center is 0.30. If the other circumstances are the same as in the previous example, the standard error of one ground coordinate is $\pm$ 22 mm. The relative error with regard to the distance 3 km from the nearest control point is $8 \cdot 10^{-6}$. According to our formula (3.70), this corresponds to "class number" $M = 2.3$.

Bibliography

Arley, N; and K.R.Buch. *Introduction to the theory of probability and statistics.* John Wiley & Sons Ltd, London, 1950.

Banachiewicz, T. *Études d'analyse pratique.* Bulletin de l'Academie Polonaise des Sciences et des Lettres. Cracovie, 1938.

Banachiewicz, T. *Outline of Cracovian algorithm of the method of least squares.* Astronomical Journal, V 50, 1942.

Grossmann, W. *Grundzüge der Ausgleichungsrechnung.* Springer-Verlag Berlin/Göttingen/Heidelberg, 1961.

Hamilton, W.C. *Statistics in physical science. Estimation, hypothesis testing and least squares.* The Ronald Press Co. New York, 1964.

Hohn, F.E. *Elementary matrix algebra.* Macmillan Co. New York, 1964.

Hristow, W.K. *Grundlagen der Wahrscheinlichkeitsrechnung, mathematischen Statistik und Methode der kleinsten Quadrate.* VEB Verlag für Bauwesen Berlin, 1961.

Rainsford, H.F. *Survey adjustments and least squares.* Constable & Co Ltd, London, 1957.

Smart, W.M. *Combination of observations.* Cambridge, Great Britain, 1958.

Tienstra, J.M. *Theory of the adjustment of normally distributed observations.* N.V.Uitgeverij "Argus". Amsterdam, 1956.

Wolf, Helmut. *Ausgleichungsrechnung nach der Methode der kleinsten Quadrate.* Ferd.Dümmlers Verlag, Bonn, 1968.

INDEX